BLUE GREEN ALGAE

Survival Strategies in Diverse Environment

The Author

Dr. Siba Prasad Adhikary (b. 1954) passed M.Sc. in Botany obtaining 1st position in 1st class from Berhampur University in the year 1976 and awarded gold medal. After his initial training on nitrogen fixing blue green algae for Ph.D. at Berhampur University (1976-1980) he had worked at Post doctoral level in the Institute of Microbiology , Freiburg i.Br., Germany (1982-1983) and National Institute for Basic Biology, Okazaki, Japan (1991-1992), and as visiting fellow at the University of Konstanz, Germnay (1996), Bhabha Atomic Research Centre, Mumbai (1987), and Institute of Microbiologysed rock surfaces of temples and monuments of India and their mechanism of stress tolerance, upgradation and refinement of the BGA biofertilizer technology for use in rice cultivation, marine algae of Orissa coast, especially of the Chilika lagoon and algal blooms of temple tanks. He is maintaining a blue green algae germplasm of 192 species with strain history. Has been awarded the 1st D.Sc. degree in Botany by Utkal University in 2000. He is a Fellow of Indian Botanical Society and was a member of the executive council and editorial board of the Phycological Society of India. Published over 130 research papers in various reputed journals of India and abroad. Awarded the Young scientist medal by Indian Botanical Society in 1983 , Bijoy Gobinda Memorial Scientist award in Life sciences for 1999, Samanta Chandra Sekhar award in Life Sciences for 1999 by the Orissa Bigyan Academy and the prestigious UGC Research Award under IX th paln (1999-2002). Thirteen students have obtained Ph.D. degree under his guidance. Written several science articles in Oriya language for popularisation of science. Operated research projects supported by UGC, CSIR, Ministry of Environment and Forests and Department of Biotechnology, Govt. of India. Presently working as a Reader in the P.G.Department of Botany as well as in the P.G.Department of Biotechnology (DBT supported M.Sc. teaching programme) of Utkal University, Bhubaneswar, operating a CSIR project on marine algae of Orissa and their bioactive substances, and also collaborating in an AICOPTAX- all India coordinated project on freshwater algae supported by Ministry of Environment and Forests, Govt. of India.

BLUE GREEN ALGAE

Survival Strategies in Diverse Environment

Dr. Siba Prasad Adhikary

P.G. Department of Botany & Biotechnology
Utkal University, Bhubaneswar (Orissa)

Pointer Publishers

Jaipur 302 003 (Raj) India

Distributed by
Prem C. Bakliwal
Aavishkar Publishers, Distributors
807, Vyas Building, Chaura Rasta
Jaipur 302 003 (Raj) India
Telefax : 2578159
e-mail : aavishkarbooks@hotmail.com

ISBN 81-7132-441-X

First Published in 2006 by
Mrs. Shashi Jain for
Pointer Publishers
Vyas Building, S.M.S. Highway
Jaipur 302 003 (Raj) India
Phones : 2578159, 2708286
e-mail : pointerpub@hotmail.com
web site : www.pointerpublishers.com

Typesetting by
Pointer Computers
Jaipur 302 017 (Raj) India

Printed at
Sheetal Printers
Jaipur 302 003 (Raj) India

Preface

A thrust area of contemporary biological research deals with tolerance to environmental stresses. As some of the earliest forms that originated on this planet nearly 3 billion years ago, blue green algae (Cyanophytes, presently known as Cyanobacteria) have successfully lived through a variety of environmental abiotic stresses, such as extreme heat of the primitive earth, extreme cold of the ice ages, anaerobiosis prevalent at the time of their origin to subsequent oxygeny for which they have been responsible. Today they are almost ubiquitous in distribution, being present from the freezing waters of Antarctic to the hot springs, from fresh-water to brackish-water or oceans and deserts. Over the years they have adapted themselves to several man-made stresses like various types of industrial effluents of different chemical make-up, chemical fertilizers, pesticides, heavy metals etc. For these reasons, they are considered very appropriate model system for studies of stress and adaptive responses. The informations obtained on their adapatation to stress has a direct utility for exploitation of their potential as green manure or nitrogen biofertilizer in stressful agricultural environments. Since higher plants are not easily amenable to ecophysiological and biochemical manipulations, suitable model systems for studying the responses of plants to environmental stresses are needed. Blue green algae possessing vital metabolic processes like photosynthesis and nitrogen fixation meet the requirement of an appropriate model system for such studies.

In this book informations about occurrence of various blue green algae and their survival in various industrial effluent polluted areas have been presented. In addition, growth and nitrogen fixation of blue green algae in pesticide burdened rice field soils, epilithic blue green algae occurring on the exposed surfaces of temples and monuments of various regions of India, their mode of survival, presence of UV sun-screen pigments in their external envelope layers, and stress proteins of these organisms has been given. Ecophysiology of blue green algae inhabiting the thermal springs has been reviewed and their possible stress adapting mechanisms was compared with those of the mesophiles occurring on exposed rock surfaces at higher temperature.

The work presented in this book has been carried out while working at various colleges of Orissa (1980 to 1986) and later at the P.G.Department of Botany, Utkal University, Orissa (1987 to 1999). Many analytical techniques used in the experiments and described in this book in different chapters I have learnt while working at the Algal Physiology laboratory, Berhampur University, Orissa (1976-1980); Institüt für Mikrobiologie, Albert Ludwigs Universität, Freiburg i.Br., Germany (1982-1983); National Institute of Basic Biology, Okazaki, Japan (1991-1992), Lehrstuhl für Physiologie und Biochemie der Pflanzen, University of Konstanz, Germany (October to December 1996) and Bhabha Atomic Research Centre, Tombay, Mumbai (September to October 1997) under the supervision of Prof. H.Patnaik, Prof. J. Weckesser, Prof. Y. Fujita, Prof. P.Böger and Dr. S.K. Apte respectively. I woe them greatly for imparting this knowedge to me. I express my indebtness to several of my research colleagues and research students: Drs. Jayanti K. Sahu, A.K.Bastia, Late D.P.Satapathy, M.K.Das, B.Rath, A.Mishra, P.Tripathy, J.K.Sahu, A.Roy, H.K.Nayak, J.Rath, B.Pattanaik, Mr. P.K.Tripathy and Mr. I.C.Tripathy for their cooperation and help in many of the experiments presented in this book. I am indebted to Prof. S.N.Patnaik, Former Professor and Head of the Department of Botany, Utkal University for his interest and support for this venture.

Siba Prasad Adhikary

Contents

List of Abbreviations and Symbols

AD	Anno domini
APHA	American public health association
AR	Analytical reagent
ARA	Acetylene reduction activity
BDH	British drug house
CCAP	Cambridge culture collection algae and protozoa
Chl	Chlorophyll
Ci	Curie
Cm	Centimetre
d	day/days
DB	Dust base
E	East
EC	Effective concentration
EC_{50}	50 per cent effective concentration
e.g.	exampli gratia (for example)
et al	et albei (and others)
etc	et cetera
Fig	Figure
G	Granule
g	Gram
x g	acceleration due to gravity

g/l	gram per litre
h	hour/hours
Hsp	Heat shock protein
i.e.	that is
In situ	in nature
In vivo	in the living organism
kDa	Kilo dalton
m	metre
mg	milligram
min	minute
mm	millimetre
μ l	microlitre
μ mol	micro mole
mmol	milli mole
N	North
nm	nono metre
n mol	nano mole
pa	pascal (0.01 millibar)
PCC	Pasteur culture collection
pH	$- \log (^aH^+)$
ppm	parts per million
rpm	revolutions per minute
S.D.	Standard deviation
UU	Utkal University
UV	Ultra violet
V/V	Volume per volume
viz	Videlicet (namely)
W	Watt
W/m^2	watt per square metre
wt	weight
w/v	weight per volume
°C	degree celcius
%	per cent
±	plus or minus
>	greater than
<	less than

1

General Introduction

Blue green algae (Cyanophytes/ Cyanobacteria) represent a group of prokaryotes that have been on the earth almost from the begining of life (Schopf, 1975). They are uniquely positioned in evolutionary hierarchy as they harbour a photosynthetic apparatus similar to that of higher plant chloroplasts within a typical bacterial cell. They possess light-harvesting pigments such as chlorophyll-*a*, carotenoids and phycobiliproteins and carry out oxygenic photosynthesis using water as an electron donor. The group exhibit considerable diversity in form, habit and function. This diversity covers morphological features (unicellular, filamentous, heterocystous, sporulating forms), ecological niches and habitats (sea water, fresh water, halophiles, thermophiles, cryophiles) and modes of nutrition (obligate photoautotraphy, facultative photo- and chemo-heterotrophy or diazotrophy). For a wide range of applications as biofertilizers and as sources of single cell protein, vitamin B-12 and pigments like phycocyanin and β carotene, these organisms are used. Their cultivation in the laboratory or on a large scale is simple and by and large requires only light and mineral salt media (Venkataraman, 1969). Thus they present themselves as an interesting group for study to ecologists, morpho-geneticists, evolutionary biologists, physiologists and biochemists as well as to agricultural scientists and biotechnologists. They are regarded as model systems for studying fundamental processes such as photosynthesis, nitrogen fixation and adaptation to environmental stresses.

1.1 STUDIES ON BLUE GREEN ALGAE IN ENVIRONMENTS POLLUTED BY INDUSTRIAL EFFLUENTS

During the past few decades man has been modifying his environment in many ways and to an extent which was undermt of prior to the enormous technological advances of the twentieth century. The industrial revolution coupled with urbanization and increasing population is accelerating the pace of these modifications. By ignoring the fundamental laws of nature, the use of environment, commonly governed by immediate expediency has made our environment highly contaminated. Water is one of the man's most precious and important natural resource and if we continue to destroy our natural living environment by polluting them the survival of human race will be in jeopardy.

The pollution of water resources results due to the discharge into them of domestic, industrial, mining, agricultural, radioactive and many other wastes. Addition of such wastes greatly alter the aquatic ecosystem. Although aquatic ecosystems have the ability to assimilate a certain amount of waste materials, the constant use and reuse of water from the natural aquatic systems by industries, agriculture sectors, municipalities and other purposes, reduce this assimilatory power. Unmanged use of certain chemicals like DDT, polychlorinated compounds and heavy metals produces a wide variety of effects when they are discharged into water. These chemical pollutants, being not easily degraded, accumulate in food chains.

The addition of various kinds of wastes to any water body renders the water unfit for public use and for the indigenous aquatic organisms. The addition of sewage, agricultural wastes and most organic industrial wastes increases the nutrient value of natural waters. The term eutrophication is generally used for the increased nutrient status of the water bodies. In developing countries like India, this problem becomes much more evident because of urban drainage, agricultural run-off, sewage, effluents and industrial wastes which are either partially treated or not treated at all. These excessive nutrients deteriorate the water quality due to excessive growth of microorganisms, algal blooms and other aquatic plants, and eventually disturb the aquatic ecosystem.

After receiving the effluent there is a drastic change in the physico-chemical and biological life of the receiving water bodies. Since aquatic ecosystems are the first to experience the effect of effluent, a study involving change in their ecology and physiology would be of paramount importance. Different habitats having similar conditions and resources are often occupied by recognizably similar biological communities. When this natural environment is upset by the addition of pollutants, only the organisms tolerant to these pollutants will survive and multiply whereas the sensitive or less tolerant forms will disappear or become restricted. The type of organism destroyed and the extent of destruction reflect the characteristics and quality of the waste entering the habitat. On the above logic, therefore, many biologists believe that organisms living in such habitats provide a fairly sensitive and reliable measure of the change of conditions. During the last 25 years the studies carried out in various laboratories of India involving changes in aquatic biota as a means of estimating water pollution and also algal bioassays to evaluate the pollution status have been summarized in Table-1. Algae, being primary producers and occurring widely in almost all aquatic habitats including polluted waters, can if suitably employed, serve as indicators of habitat conditions (Palmer, 1969). In recent years these microorganisms have been widely used as test systems for evaluating the extent of pollution caused by industrial wastes (Fitzgerald, 1971; Rai and Kumar, 1976 a,b; Gour and Kumar, 1986; Manoharan and Subramanian, 1993).

Table 1: Summary of the research work carriedout at various laboratories of India on taxonomical-ecological studies of algae and the effect of industrial effluent of polluted habitats on their growth

Name and location of the Industry	Organism used in the study	Type of study	Reference
1	2	3	4
Zawar Mines, Rajastan Zinc smelter, Rajastan	*Plectonema boryanum, Anabaena doliolum, Fritschenella mucicola, Tolypothrix tenuis, Calothrix brevissima*	Effect of industrial wastes on growth of these organisms	Rana *et al.*, 1971
Indian oil refinery, Barauni; Bihar fertilizer factory, Sindri; Mohan Meakin brewery, Ghaziabad	*Chlorella* sp. *Anacystis nidulans*	Physico-chemical characteristics of effluent, taxonomical-ecological study of algal flora of the polluted habitat and uptake of nutrient by algae in the study sites	Kumar and Sharma, 1974
Fertilizer factory, Sahaupuri, Varanasi	(i) *Chlorella* sp	Evaluation of nutrient status of the effluent using algae	Rai and Kumar, 1976 a
	(ii) *Chlorella vulgaris Aracyslis nidulaus*	Nutrient uptake and growth of test organisms in presence of industrial effluents	Rai and Kumar, 1976 b
Kanoria chemical factory, Renukot; Rohtas paper industry, Dalmianagar	*Chlorella* sp., *Anacystis nidulaus*	Physico-chemical analysis of effluent and effect of industrial effluents on growth of the organisms	Agarwal and Kumar, 1978

Contd. ...

1	2	3	4
Mandya national paper mills, Mysore	Algal indicator species occurring in effluent polluted habitat	Ecological study (physico-chemical analysis of effluent and algal flora of the effluent polluted area)	Somashekar and Ramaswamy, 1983
Oil refinery (Hindustan Petroleum Ltd., Gangapuram, Andhra Pradesh)	*Scenedesmus incrassatulus Synechococcus aeruginosus*	Physiological responses of the algae to oil refinery effluents	Reddy *et al.*, 1983
Sikkim distilleries, Rangpo	*Chlorella vulgaris*	Ecophysiological study (Physico-chemical analysis of effluent, algal flora and tolerance of *Chlorella* to industrial effluents)	Venu *et al.*, 1984
Fertilizer factory, Punjab	*Anabaena variabilis, Chlorella vulgaris*	Physiological responses of a green alga and a blue green alga to fertilizer factory effluent; green algae was more tolerant	Ahluwalia and Arora, 1986
Assam oil refinery, Gauhati	*Selenastrum capricormutum*	Testing the alga to evaluate the potentiality of oil refinery effluents to sustain algal growth	Gour and Kumar, 1986
Paper mill, Sambalpur, Orissa	Phytoplankton productivity studies in effluent polluted habitat	Phytoplankton population and productivity in paper mill effluent receiving waters of river Ib.	Kar *et al.*, 1987
Assam crude oil; Oil refinery, Gauhati	*Anabaena doliolum*	Effect of crude oil on photosynthesis and associated electron transport of the alga	Singh and Gour, 1988
National fertilizer complex, Nangal; Steel manufacturing plant, Ludhiana; Automobile factory,	*Scenedesmus* sp.	Physical-chemical characteristics of effluent of the industries and their effect on chlorophyll-a content of the alga	Ahluwalia *et al.*, 1989

Contd. ...

1	2	3	4
Ludhiana and Electroplating plant, Chandigarh Petroleum oil (furnase oil, petrol, Kerosene, crude oil) and waste oil from different industries	*Anabaena doliolum*	Effect of various petroleum oil on photosynthetic ETS of the blue green alga	Singh and Kumar, 1991
Asia Glues and chemicals Ltd., Madurai (Ossein effluent)	*Oscillatoria psendogerminata* var. *unigranulata*	Removal of nitrate, ammonia and phosphate from effluent by the blue green alga was observed. Influence of effluent on its photosynthesis and nitrogen metabolism was studied.	Manoharan and Subramanian, 1993
Star paper mill and Board mill, Saharanpur	*Scenedesmus* sp.	Evaluation of nutrient status of the effluent using algal bioassay; lower concentrations supported growth of the alga	Kaur *et al.*, 1993
Orient Paper Mill, Brajarajnagar, Orissa	*Westiellopsis prolifica*	Chlorophyll, carotenoid and protein content of the blue green alga increased in presence of paper mill waste water. Low cost biological treatment using *W. prolifica* through nutrient manipulation has been suggested.	Dash and Mishra, 1996

Growth of photosynthetic microorganisms also leads to fall in the concentration of waste products like nitrate and phosphate and thus helps in complete or partial purification of waste water. Several authors have emphasized the importance of waste grown algae in recirculation of waste water for industrial purpose (Fogg, 1971; Oswald and Golueke, 1971; Zajic and Chiu, 1970). However, little attention has so far been paid to understand the ecophysiology and adaptive features of these organisms in such habitats, and to assess their role in retrieving the pollutants consequently decreasing the pollution load of the water bodies.

1.2 STUDIES ON THE INTERACTION OF BLUE GREEN ALGAE WITH PESTICIDES IN THE RICE FIELD ENVIRONMENT

Rice is the staple food for atleast half of the world's population (Swaminathan, 1984). The Asian rice area is 90 per cent of the global total, in a region which supports >60 per cent of the world's population. As in the developing countries the nitrogenous fertilizer consumption per hectare of agricultural land is much less (an average of 10 kg N/hectare), nitrogen fixing blue green algae contribute significantly to the nitrogen economy of rice crop of the region (Roger and Kulasooriya, 1980). Blue green algae multiply rapidly in the flood water in neutral to alkaline soils and release nitrogen slowly for meeting the nitrogen requirement of rice (Triol *et al.*, 1982; Whitton *et al.*, 1988). Since the agronomic potential of these organisms was recognised by De (1939), long-term fertility experiments and nitrogen-fixation measurements have confirmed the importance of blue green algae in maintaining a moderate but constant rice production in fields receiving no fertilizer (Roger, 1988) and many trials have been conducted to increase rice yield by their inoculation in the fields (Venkataraman, 1978, 1981).

In man's continued effort to enhance food production, new high yielding crop varieties and new techniques for crop and field management are being constantly evolved resulting in a parallel increase in occurrence of pests which calls for their efficient management. Pest control practices were further developed with advances in science and technology and today pesticides are used on an extensive scale in agricultural fields. But the emphasis on chemical control of agriculture pests have caused a serious imbalance in the agro- ecosystem. Since about 70 per cent of the amount of the chemicals sprayed on the field crops do not stick to the plants, the enormous quantities of chemicals that fall on the earth get mixed up well with soil and become poisonous to microbial life. The non-target effect of pesticides on nitrogen fixing blue green algae of rice fields is considerable considering the role of these organisms in maintenance of sustainability of rice production especially in India where rice is the principal crop of millions living in its villages. This reflects a justifiable concern for the fate of nitrogen fixing blue green algae in pesticide burdened paddy fields (Mishra *et al.*, 1989)since according to Watanabe (1962) and many others (Venkataraman, 1981; Roger and Kulasooriya, 1980) the average amount of nitrogen fixed by these organisms under normal conditions in a flooded paddy field is 25-30 kg hectare^{-1} year^{-1}.

Several studies examining response of blue green algae to different pesticides have been published during the last three decades. Summary of these studies are presented in Table-2.

Table 2 : Summary of the studies examining the response (s) of blue green algae to pesticides

Sl no.	Organism	Insecticide / herbicide / fungicide tested	Concentration	Findings	Reference
1	2	3	4	5	6
1.	28 species of blue green algae	Ceresin, Dithane, 2,4—D, Delapon, Propazine, cotoron, Diuron, Linuron	0.0-0.34 kg/ha 1.7-16.8 kg/ha 1.4-4.5 kg/ha 5.6-56.0 kg/ha 2.24 kg/ha 1.12-2.24 kg/ha 0.6-12.1 kg/ha 0.6-6.9 kg/ha	*Tolypothrix tenuis* grew in presence of all the pesticide concentrations except of Diuron. *Anlosira fertilissima* tolerating higher levels of Ceresan. Dithane, 2.4-D and Delapen wassensitive to Cotoron, Diuron, Linuronand Propazine. Maximum tolerancewas exhibited by *Anacystis nidulans*	Venkataraman and Rajyalaxmi 1971
2.	27 species of blue green algae	Ceresin, Dithane, Delapon	0.1 to 100 ppm	Most of the species of *Anabaena* tolerated 100 ppm of Ceresan. Dithane was lethal to some species of *Anabaena* and *Nostoc* even at the lowest concentrations. At 100 ppm of Delapon	Venkataraman and Rajyalaxmi 1972

Contd. ...

1	2	3	4	5	6
				almost all species grew well.	
3.	*Aulosira fertilissima*	BHC, Endrin, Diazinon, Sevin, Propanil	10 ppm	BHC stimulated growth while Endrin suppressed the growth of the organism in rice fields. Same concentration of Diazinon, Sevin and Propanil had no inhibitory effect on the blue green alga.	Ahmed and Venkataraman, 1973
4.	*Cylindrospermum* sp. *Aulosira fertilissima* *Plectonema boryanum*	BHC Lindane Diazinon Endrin	10 to 500 ppm	BHC was more toxic among all the pesticides tested. A *fertilissima* and *P.boryanum* were more resistant than*Cylindrospermum* sp.	Singh. 1973
5.	8 different species of blue green algae	Diquat Paraquat Malathion	25-200 porn	Dlaquat and Paraquat were lethal to allthe blue green algae at 25 ppm. Same concentration of Malathion had partially Inhibited growth of *Chlorogloea fritschii* but at 200 ppm completely inhibited its growth.	Da Salva *et al.*, 1975
6.	*Nostoc muscorum*	Carbofuran (Furadan. 3% G)	25-1000 µg / ml	Higher concentrations (100-500 µg. ml) of the pesticide was more toxic at the acidic pH and the	Kar and Singh 1977

Contd. ...

1	2	3	4	5	6
				toxicity reduced at the alkaline pH of the medium. Reduction in toxicity was observed at higher light intensity and also due to higher population density in the culture	
7.	*Gloeocapsa alpicola* *Anabaena cilindrica* *Anabaena variabilis* *Nostoc mucorum* *Nosroc entophytum* *Tolypothrix tenuis*	Propanil 3,4-Dichioro aniline	0.01-5 ppm	Higher concentration of Propanil (> 5 ppm) inhibited the growth and photosynthesis of all the organisms. 3-4-Dichloro aniline was less toxic than Propanil	Wright *et al.* 1977
8.	*Nostoc muscorum*	Carbofuran (Furadan, 3% G)	25-1200µg / ml	Growth remained unaffected at 10 µg/ml while at 25 µg/ml concentration growth was stimulated. 1200 µg/ml was sub-lethal dose	Kar and Singh, 1978
9.	*Nostoc muscorum*	Machete	10-20 µg/ml	More than 20 µg/ml of the herbicide was toxic to the alga	Singh and Vaishampayan, 1978
10.	*Nostoc muscorum* *Wollea bhardwajae*	Carbofuran HCH	4-30 µg / ml 4 µg / ml	Toxicity of both the insecticides could be removed by repeated cultivation of the alga	Kar and Singh, 1979 a

Contd. ...

1	2	3	4	5	6
11.	*Nostoc musorum*	Hexachloro cyclo hexane (HCH)	4 ppm	Toxicity of HCH was reduced by increasing the concentration of thenutrients K_2HPO_4 (10 to 40 ppm andCaCl$_2$ (55 to 330 ppm) in the Culture medium	Kar and Singh, 1979b
12.	*Westiellopsis* sp. *Aulosira* sp. *Tolypthrix* sp. *Nostoc* sp.	Thimet	1-1000 ppm	The pesticide did not exhibit a harmful effect on *Westiellopsis* sp., *Aulosira* sp. and *Tolypothrix* sp. at the level of 1000, 500 and 300 ppm respectively, while *Nostoc* sp. could not tolerate even 1 ppm	Gangawane, 1979
13.	*Calothrix* sp. *Westiellopsis* sp. *Aulosira* sp. *Tolypothrix* sp. *Nostoc* sp.	Topison-M	50-1000 ppm	50 ppm of Topison-M reduced the growth of *Calothrix* sp. by 11.5%, 300ppm of the same pesticide reduced the growth of *Aulosira* sp. and *Tolypothrix* sp. by 12.8% and 14% respectively. Its 1000 ppm reduced the growth of *Westiellopsis* sp. and *Nostoc* sp. by 31.6% and 22.5% respectively	Gangawane and Kulkarni, 1979

Contd. ...

1	2	3	4	5	6
14.	*Calothrix* sp. *Westiellopsis* sp. *Aulosira* sp. *Tolypothrix* sp. *Nostoc* sp.	Captafol	100-1000 ppm	*Nostoc* sp. Tolerated the highest concentration of captafol (1000ppm) followed by *Tolypothrix* sp. (500ppm), *Aulosira* sp.(300ppm) and *Calothrix* sp(100ppm)	Gangawane and Saler, 1979
15.	*Anabaena vaginicola* *Cylinarospermum muscicola* *Nostoc entophytum*	Dimecron Ekalux Blitox Dithane	5-20 ppm	Higher concentration of Dimecron was toxic to all these organisms. Higher concentrations of Ekalux (5,10,20ppm) had immediate deleterious effect; 8-10 ppm of Blitox and Dithane were lethal to all the blue green algae.	Anand and Veerappan, 1980
16.	*Mastigocladus laminosus* *Tolypothrix tenuis*	Tolkan Fluchloralin	50-500 ppm	*M. laminosus* tolerated both the pesticides up to 100 ppm, where as *T. tenuis* tolerated Tolkan and Fluchloralin up to 500 and 50 ppm respectively	Khalil et al. 1980
17.	*Anabaena doliolum*	Butachlor	0.05 to 20 ppm	20 ppm was lethal, sub-lethal dose was 5 ppm which caused	Kashyap and Pandey 1982

Contd.....

1	2	3	4	5	6
				a decline in protein, phycobilin levels, inhibition of heterocyst differentiation and nitrogen fixation	
18.	*Hapalosiphon fontinalis Hapalosiphon welwitschii Westiellopsis prolifica Calothrix braunii*	BHC Phorate	1,10,100 ppm	BHC and Phorate at 100 ppm had inhibitory effect. Carbofuran above 1 ppm was inhibitory. At 10 ppm level nitrogenase activity was reduced by 25% with Phorate, 50% with BHC and 87% with Carbofuran	Kaushik and Venkataraman 1983
19.	*Anabaena* sp.	Sevin	10-1500 µg/ml	1000 µg/ml was sub-lethal dose for Anabaena sp. while 1500 µg/ml was lethal for both the blue green algae. Single or two successive additions of Nuvacron to soil had no inhibitory effect on blue green algal flora only when applied close to field dose (0.5 to 2 kg/ha)	Adhikary *et al.* 1984, Megharaj et al. 1986
20.	*Westiellopsis prolifica* Blue green algae in field	Nuvacron Ekalux	0.5 to 5 kg/ha		
21.	*Anabaena doliolum* *Nostoc muscorum*	Butachlor	2.5-20 ppm	*Anabaena doliolum* was found to be more tolerant than either	Pandey and Kashyap, 1985

Contd. ...

1	2	3	4	5	6
	Anacystis nidulans			*Nastoc muscorum* or *Anacystis nidulans* to Butachlor; lethal concentrations being 20, 5 and 2.5 ppm respectively	
22.	*Tolyootnrix ceylonica* *Scytonema* *cincinnatum*	Butachlor Stam F 34	1-100 ppm	Growth and chlorophyll synthesis were not affected up to 10 ppm; a maxed inhibition was observed In both the blue green algae.	Roychoudhury and Kaushik 1986
23.	*Anacystis nidulans*	Demeton Sevin Difolaton Dithane M-45	1-500µg/ml	Final tolerance levels of *Anacytis nidulans* to Demeton, Sevin, Difolatonand Dithane M-45 were 500, 1, 100 and 100µg/ml respectively	Thomas and Shanmugasundaram, 1986
24.	Blue green algae in field	Cypermethrin, Fenvalerate	5-50 ppm	Stimulation of growth of blue green algae was observed at 5 ppm in the rice fields; stimulation or inhibition of their growth varied between 10-50 ppm depending on organism	Megharaj *et al.*, 1987
25.	*Anabaena* ARM 286 *Anabaena* ARM 310	BHC Ekalux	1-10 ppm	10 ppm of BHC was more stimulating for oxygen evolution	Subramaniam *et al.*, 1987

Contd. ...

1	2	3	4	5	6
				than 1 ppm. The increase of photosynthetic rate with Ekalux was relatively higher than BHC.	
26.	9 species of blue green algae of rice fields	Carbofuran (Furadan 3%G)	0.5-5 kg/ha	Application of 0.5 and 1 kg/ha of Furadan to soil under non-flooded conditions did not affect blue green algae population, highest level of 5 kg/ha was toxic	Megharaj *et al.*, 1988
27.	Blue green algae of rice fields	Butachlor Oxadizon Bethiocarb Pendimethalin	0.5-1.5 kg/ha	Butachlor and Oxidiazon resulted in higher toxicity to blue green algae than Benthio-carb and Pendimethalin.	Singh *et al.* 1988
28.	*Nostoc muscorum Gloeocapsa* sp.	Butachlor Fluchloralin Propanil	Up to 100 µg/ml	Semi-lethal doses of Butachlor and Fuchloralin showed little affect on photosynthetic oxygen evolution of both the species whereas Propanil showed inhib-itory effect at the same con-centration. Respiratory oxygen uptake showed stimu-latory effect in presence of all the three herbicides	Singh and Tiwari, 1988

Contd. ...

1	2	3	4	5	6
29.	*Anabaena* sp. *Aulosira fertilissima*	Malathion Endosulfan	1-500 ppm	Malathion inhibitd growth of *Aulosira fetilissima* at > 10 ppm but *Anabaena* sp. could tolerate up to 500 ppm. Even 1 ppm of Endosulfan was inhibitory for both the blue green algae.	Tandon *et al.*, 1988
30.	*Anabaena variabilis* *Aulosira fertilissima* *Scytonema chiastum* *scytonema stuposum*	Bavistin Eenlate Captan Dithane Cyathion	100 to 400 ppm	Lethal effect was observed in *Anabaena variabilis* at 100ppm of Captan, 400ppm of Cyathion, in *Aulosira fertilissima* at 400ppm of Dithane, 200 ppm Cyathion, in *Scytonema chiastum* at 100 ppm of Bavistin, 400 ppm of Cyathion and in *Scytonema stuposum* at 400 ppm of Bavistin, Beniate , Dithane and 200 ppm of Capton and Cyathion.	Dikshit and Tiwari, 1992
31.	*Anabaena khannae* *Calothrix marchica* *Nostoc calcicola* *Tolypothrix limbata*	Butachlor Benthiocarb Pandimethalin Oxadiazon	0.5 to 1.5 ppm more resistant.	*Anabaena khannae* and *Calothrix marchica* were proved to be	Kolte and Goyal, 1992
32.	*Anabaena oryzae* *Anabaena anomala*	Carbaryl BHC	0.5 to 10 µg/ml	Increasing concentrations of these weedicides from 0.5 to	Nagpal and Goyal, 1992

Contd. ...

1	2	3	4	5	6
	Nostoc paludosum *Nostoc calcicola*			50 ppm resulted in progressive reduction in growth and chlorophyll-a content of all the blue green algae	
33.	*Anabaena variabilis* ARM 140 and ARM 143	Carbaryl BHC	0.5 to 10 µg/ml	10 µg/ml of Carbaryl as well as BHC was lethal for both the strains of *Anabaena variabilis* ; ARM 143 was relatively more sensitive to the pesticides	Pabbi and Vaishya, 1992
34.	*Anabaena oscillarioides* *Nostoc khilmani*	Butachlor	0.1-5 pp,	0.1 and 1ppm concentrations of Butachlor stimulated pigment content and nitrogenase activity though higher concentrations was inhibitory. *Nostoc khilmani* was more tolerant than *A. oscillarioides*.	Soliman *et al* , 1993
35.	25 species of rice field blue green algae	Furadan (3%G), commercial grade	10-80 mg/ml	Among the test orgamsms, all the species of *Calothrix* tolerated up to 70 mg/ml of Furadon while *Aulosira* sp. was most sensitive. Sheathed forms of blue green algae were more tolerant.	Rath and Adhikary, 1994

Contd. ...

1	2	3	4	5	6
36.	*Aulosira fertilissima* ARM 68 *Nostoc muscorum* ARM 221	Monocrotophos Malathion Dichlorovos Phosphomidon Quinolphos	1-250 ppm	The optimal concentration for growth of monocrotophos, Malathion, Dichlorovos, phosphomdon and Quinolphos was 100, 75, 25 and 1 ppm respectively. Both the species grew maximally with the pesticides in the absence of inorganic phosphate suggesting their utilization as the sole source of phosphorus.	Subramaniam *et al.*, 1994
37	*Anabaena fertilissima* *Anabaena variabilis*	Furadan 75DB (analytical grade)	100-250 µg/ml	PH, Irradiance and population sizes modified the toxic effect of Furadan; toxicity was increased at acidic pH, low irradiance and with lower population size in culture.	Rath and Adhikary, 1996
38	10 species heterocystous blue green algae	Sevin Rogor Hildan	0.1-500 ppm	Among the species tested, *Calotnrix parietina* UU 1423 and Calothrix sp. UU2427 possessing a well defined sheath were more tolerant to all the three pesticides.	Das and Adhikary, 1996

These works relates to the studies on toxicity of several insecticides, herbicides, and fungicides to blue green algae in culture, mostly on finding out the tolerance limits and on accumulation and metabolism of the agrochemicals. However, little emphasis has been paid to identify and isolate blue green algal forms capable of tolerating relatively higher concentrations of the pesticides and studying the adaptive features of these organisms which enable them to thrive in pesticide burdened soils. Since now-a-days blue green algal biofertilizer containing selected strains are inoculated into paddy fields along with reduced dose of commercial fertilizers to maintain the soil health, it is essential to use species which have greater capability to survive the agrochemical stresses, i.e. those could thrive in presence of recommended or even higher dose of the agrochemicals under rice field environments without decreasing their nitrogen fixing ability.

1.3 STUDIES ON BLUE GREEN ALGAE ADAPTED TO TEMPERATURE AND DESICCATION STRESSES

1.3.1 Terrestrial blue green algae

The blue green algal forms, although generally occur in freshwater or marine habitats, also occupy a variety of terrestrial environment where their ecological importance is considerable. Most commonly they occur either on the surface or at a depth of up to several centimetres in soil, live in the crevices and surface of rocks, inhabit snow, tree trunks and fallow rice fields. Most of these terrestrial habitats represent extreme environments characterized by aridity and temperature and light extremes. In addition, in many of these habitats, frequent fluctuation of environmental conditions occur in sharp contrast with the mostly stable aquatic habitats. To cope with these stress situations, terrestrial blue green algae develop morphological and physiological adaptations or occupy sheltered microhabitats making them successful even at these extreme habitats. Desiccation tolerant cells in algae include zygotes, akinates, cysts and spores. Especially in blue green algae many have unmodified vegetative cells that are resistant to desiccation. The classical studies of Fritsch (Fritsch, 1922; Fritsch and Hains, 1923) and of Stokes (1940) remain as the most direct ecological examinations of drought tolerant organisms. One of the main conclusions of their studies was that the water status of the soil is of paramount importance for the growth and survival of algae.

Blue green algae seem to have an overall better survival rate than green algae and quite often they are pioneering organisms (Bewley, 1979). The 'floristic' composition of desiccated soils is related to the dryness of the biotope. In a paddy field in Italy, where the period of drought is relatively short, only 30% of the algal flora were blue green algae, whereas in of Senegal, where the dry season lasts about 8 months, they constituted more than 90% of the soil algal flora (Roger and Reynand, 1982). The environmental factors which control their population in soil seems to be light, humidity, temperature, availability of nutrients and pH. In fallow soils the density of these organisms was generally more in the upper few centimetres and falls off rapidly with depth (Tomaselli and Giovannetti, 1993). About 10-15 cm top layer of the soil is invariably exposed to extreme variations in the moisture content. This makes most of the edaphic algae capable of surviving long spells of drought stress and desiccation. Seasonal variation in the blue green algal population of soils are generally quantitative, due to fluctuations in the availability of water while the species composition remains constant over the year (Metting, 1981). Some desiccation tolerant organisms like *Cylindrospermum, Anabaena, Nostoc* and *Nodularia* form resting spores while certain others tolerate desiccation due to special morphological adaptations (Bewley, 1979).

Desiccation-tolerant blue green algae have been found in a diverse range of terrestrial environments e.g. inside the rocks in the Antarctic (Friedmann and Ocampo- Friedmann, 1984; Vestal, 1988), in subalpine soil-crust (Fritz-Sheridan, 1988), in the uppermost millimeters of sand (Graebner, 1910; Forster and Nicolson, 1981; Van den Ancker *et al.*, 1985), in soils and in rocks of hot deserts (Friedmann *et al.*, 1967; Friedmann, 1971; Levi *et al.*, 1981; Roger and Reynand, 1982; Danin and Garty, 1983), in tropical grass land (Dulien *et al.*, 1977), on the walls and roof tops of buildings (Tripathy and Talpasayi, 1980; Chada *et al.*, 1980) and as epiliths on the exposed rock surface of several monuments, archaeological remains and works of art throughout the globe (Wee, 1980; Krumbein, 1988; Ortega-Calvo *et al.*, 1993; Albertano, 1993; Albertano *et al.*, 1994; Anagnostidis *et al.*, 1983).

Epilithic blue green algae inhabit the exposed rock surfaces as a crust forming up to few mm thick layer. Blue green algae followed by green algae form the major component of the epilithic vegetation. Epilithic blue green algae inhabiting surfaces as well as crevices of rocks and stone receive moisture mainly from the atmosphere. Light intensity and humidity are the most important factors that determine the composition of an epilithic vegetation. *Chroococcus, Gloeocapsa, Stigonema, Calothrix* and *Tolypothrix* are of common occurrence on rock surface and form a blackish mass (Round, 1973). On the monuments made of sandstone, lime stone or granite, several blue green algal species occur as epilithophytes, especially on the surface exposed to sunlight and moisture (Krumbein and Lange, 1978; Ortega-Calvo *et al.*, 1994). In Jerusalem, tomb stones and cut rocks of monolith tombs of up to 2,600 years old have been found to be inhabited by colonies of unicellular blue green algae (Danin, 1983). In India, walls of old temples get disfigured due to growth of *Tolypothrix, Nostoc, Calothrix,*

Plectonema, Lyngbya, Xenococcus, Gloeothece and *Gloeocapsopsis* (Tripathy *et al.*, 1977; Roy *et al.*, 1977; Tripathy *et al.* 1999; Pattanaik and Adhikary 2002). *Tolypothrix byssoidea* showed widespread occurrence on the walls of almost all the temples of coastal region of Orissa state. It forms blackish crust during summer months which becomes bluish-green tuft soon after receiving water on the onset of monsoon. Treatment of the blue green algal crusts with 2,3,5-triphenyl-tetrazolium chloride showed that reduced activity was maintained in the cells even under extreme arid conditions (Adhikary and Satapathy, 1986). Lithophytic blue green algae are exposed to a wide range of light intensities varying from bright sunlight for epilithic forms to very low light intensities for endolithic and hypolithic forms. These blue green algal species which are exposed to strong light often have coloured sheath material. Production of lamellated sheath reduces loss of water from their cells.

1.3.1.1 Ecophysiology of blue green algae in terrestrial environments

The terrestrial blue green algae, which exist and proliferate in extreme habitats, posses the ability to withstand extended periods of desiccation and have the capacity to revive from the air-day state (Brock, 1975; Whitton *et al.*, 1979). The cosmopolitan terrestrial blue green alga, *Nostoc commune* was found to tolerate acute water stress and survive in the air dry state for many years. Under natural desiccating conditions, the cells are embedded and immobilized in a water absorbing sheath composed of carbohydrates. The blue green alga accumulates a novel group of acidic proteins when subjected to repeated drying and rehydration. These proteins occur in high concentrations and had a protective function at the structural level (Scherer and Potts, 1989). It has been shown in field material of *Nostoc commune*, as well as in immobilized axenic culture of *Nostoc commune* UTEX 584, that there was a reproducible sequential recovery of metabolic activities upon rewetting of cells begining with respiration through photosynthesis and finally nitrogen fixation (Scherer *et al.*, 1986). Respiration and photosynthesis preceeded growth and were exhibited by the existing vegetative cells, whereas recovery of nitrogen fixation depended on newly differentiated heterocysts. The first sensitive site affected by desiccation was electron transport between plastoquinone and P 700 which showed recovery on rehydration. In contrast, desiccation intolerant species did not recover when rehydrated following even limited water loss (Wilkens *et al.*, 1978) suggesting irreversible chloroplast damage. It has been suggested that the accumulation of fats and oils in the protoplasm of some autotrophic microorganisms increases their capacity to withstand desiccation (Collyer and Fogg, 1955). The mechanism of desiccation tolerance in *Nostoc commune* was found out by demonstrating that both phospholipid bilayers and proteins can be solubilized during water stress by sugars, especially trehalose (Crowe *et al.*, 1987). Trehalose was found in a variety of microorganisms including blue green algae when they were subjected to drying or osmotic stress.

As was stated before by Fritsch (1922), Fritsch and Hains (1923) and Potts and Friedmann (1981), the water status of the substratum is of foremost importance for physiological activity of terrestrial blue green algae. This implies that the nano-climate of the surface and the upper few millimetres below the surface is a controlling factor in the pattern of microbial colonization. The conditions for controlling this climate are determined mainly by temperature, solar radiation, and physical and chemical properties of the substratum. Availability of water is aided by the climatological factors and is of critical significance. Growth of non-aquatic organisms has been related to the 'cumulative imbibition time' (Danin and Garty, 1983), which is the time, the organisms are in contact with water. This factor can be improved by the water-holding properties of the substratum. The morphology of the organism itself or the direct physical environment is critical in this respect. Most of the desiccation tolerant species of blue green algae possess a sheath layer or produce mucilage around their cell wall which have great significance as both these structures could retard water loss from the cells in a drying environment (Campbell, 1979, Bewley, 1979; Scherer *et al.*, 1984; Shephard, 1987).

1.3.1.2 UV-photoprotection in terrestrial blue green algae

Blue green algae inhabiting terrestrial environments are often subjected to intense solar radiation. Under such conditions they must possess efficient mechanisms to prevent or to counteract the harmful effects of radiation in the ultraviolet region of the spectrum. Based on wavelength, ultraviolet radiation is subdivided into three regions, namely UV-A (315-400 nm), UV-B (280-315 nm) and UV-C (< 280 nm). There is an extensive literature on the effects of UV on blue green algae (Whitton, 1987). But its value for predicting the ability for organisms to survive in nature is not always clear since many of the earlier studies used UV-C radiation though this does not penetrate the earth's atmosphere. Interest in the impact of UV-B and UV-A radiations has, however, been stimulated in recent years by the apparent decrease in the stratospheric ozone shield. UV-B radiation was found to be inhibitory for several physiological process particularly photosynthesis (Tyagi *et al.*, 1991; Sinha *et al.*, 1995). Mechanisms that counteract UV-mediated photodamage in blue green algae include both dark DNA repair and photorepair (Levine and Thiel, 1987), the presence of detoxifying enzymes (Miltler and Tel-or, 1991) and action of photo-protective carotenoids (Buckley and Houghton, 1976). Among possible mechanism to prevent UV-photodamage were negative photomovement to prevent UV-photodamage and synthesis of UV-sunscreen compounds (Gracia-Pichel and Castenholz, 1991, 1993).

In the last decade presence of UV-sunscreen pigments in a variety of blue green algae has been detected. Structure of these sunscreen pigments was determined and their photo-protective role has been proposed. Summary of these findings are presented in Table-3. Presence of UV- A/B-absorbing pigment with maxima at 312 and 330 nm was first described by Scherer *et al.* (1988) from field materials of *Nostoc commune.* Subsequently, a variety of colourless, water soluble compounds known as mycosporine amino acids (MAAs) like substances that had been

Table 3: Summary of the studies so far published on the UV-sunscreen pigments in bluegreen algae

Sl No.	Subject of the study	Reference
1	UV-A/B protecting pigment in terrestrial blue green alga, *Nostoc commune*	Scherer, Chen and Böger, 1988
2	Characterization of scytonemin, a blue green algal sheath pigment	Garcia-pichel and Castenholz, 1991
3	UV-sunscreen role of scytonemin in *Chlorogloeopsis*	Garcia-pichel, Sherry and Castenholz, 1992
4	Structure of scytonemin, an UV sunscreen pigment from the sheaths of blue green algae	Proteau, Gerwick, Garcia-Pichel and Castenholz, 1992
5	Occurrence of UV-absorbing, mycosporine like compounds in blue green algae	Garcia-Pichel and Castenholz, 1993
6	UV-sunscreen role of mycosporine-like compounds in *Gloeocapsa* sp.	Garcia-Pichel, Wingard and Castenholz, 1993
7	Structure of novel-oligosaccharide-mycosporine amino acid-like UV A/B sunscreen pigment from *Nostoc commune*	Böhm, Pfleiderer, Böger and Scherer, 1995
8	Carotenoids and mycosporine-like amino acid compounds in members of the genus *Microcoleus*	Karsten and Garcia-Pichel, 1996
9	UV-B shielding role of $FeCl_3$ and certain blue green algal pigments	Kumar (A), Tyagi, Srinivas, Singh and Kumar (HD), Sinha and Häder, 1996

characterized in fungi have been detected in a number of blue green algae (Garcia-Pichel *et al.*, 1993). They represent a characteristic single absorption band in the UV spectrum between 230-400 nm, with maxima ranging from 310-360 nm and absorb only weekly below and above this main band (Garcia-Pichel and Castenholz, 1993; Böhm *et al.*, 1995; Karsten and Garcia-Pichel, 1996). In addition to the mycosporine amino acid-like compounds, a yellow-brown pigment, scytonemin, found in the extracelluar sheath of the species thriving in habitats exposed to intense solar radiation has been recently reported to play a considerable role in photo-protection (Garcia-Pichel and Castenholz, 1991). Scytonemin absorbs most strongly in the UV-A spectral region (315-400 nm with peak at 384 nm; *in vivo* max=370 nm). It has been shown that natural levels of scytonemin found in the blue green algal sheath may be sufficient to prevent 85-90 per cent of incident UV-A radiation from entering the cells (Garcia-Pichel *et al.*, 1992). Recently, scytonemin pigment has been isolated and chemically characterized (Proteau *et al.*, 1993). It is a photo-stable, dimeric molecule of Indole and phenolic subunits (mol.wt. 544) and is known only from the sheaths of blue green algae. Scytonemin formation in sheathed blue green algae thus is an adaptive strategy for photo- protection against short wavelength irradiance.

Imagine summer temperature exceeding 60° C on the exposed rock surface of temples and monuments in several regions and almost no precipitation for three consecutive months during the summer period (March-May). In these hostile environments of a tropical country like India, interesting and understudied blue green algae communities are found. They thrive on the exposed rock surface as blackish-brown crust or mat during the summer and resume their growth forming bluish-green tufts soon after receiving couple of monsoon rains. Several species of blue green algae isolated from the rock surface of the temple and monuments possessed a consistent yellowish-brown sheath and absorb strongly at the UV-A and UV-B regions of the spectrum even after cultivating them in unialgal culture under fluorescent light (Adhikary, 1996). There is a very little information available about physiological adaptations of blue green algae to such habitats. Questions arise as to whether or not these organisms possess special protective mechanisms to strong solar incidence, what levels of desiccation they can tolerate, and how they react physiologically to desiccation and high temperature stress.

1.3.2 Blue green algae occurring in thermal springs

Blue green algae are characteristic of several types of 'extreme' environments of which thermal streams are probably best known (Castenholz, 1973; Brock, 1978; Ward *et al.*, 1989). In aquatic and semi-aquatic situations, higher than ambient temperature tend to enrich blue green algal flora, usually at the expense of others. Thermal tendencies of blue green algae are exemplified in regions heated by solar and atmospheric radiations. Semi-terrestrial mats in regions of high heat, shallow ponds in hot deserts, cooling towers and ponds of industries are the examples of habitats dominated by blue green algae. In many cases it is not clear whether higher temperature or chemical factors have excluded or inhibited the usual array of eukaryotic forms.

The geothermal springs of the world sharing water temperature >48° C provide an exclusive habitat for prokaryotes (Brock, 1967). In some, blue green algae species grow at temperatures as high as 74° C, and non-photosynthetic autotrophic bacteria grow at up to 95° C (Brock *et al.*, 1971). Certain species of blue green algae occur in almost every illuminated hot springs about a pH of 5 and below 74° C in Western North America and below 64° C in much of the rest of the world (Castenholz, 1973), unless there are extreme concentrations of certain solutes. The thermal tendency of blue green algae was thought to be because of their prokaryotic structure since the reports so far available indicate that only prokaryotes are thermophiles. It has also argued that blue green algae, which are possibly the direct descendants of precambrian O_2-producing microbes, evolved in a warmer mean environment than the present day, and have retained the ability to function best in that range.

In antiquity, the presence of organisms in hot springs was noted. Plinius the Elder, in his Natural History, noted: "Green plants grow in the hot springs of Panda, frogs in those of Pisa, fishes at Vetulonia in Etruria near the sea" (1944). The hot springs at Panda (Abans) still flow, and they are colonized by blue green algae, perhaps the green plants of pliny. In the hot springs many blue green algae persist forming mat like layers throughout the year and their growth continues even during winter. The source of temperature in most hot springs seldom changes more than few degrees in an year and often not at all. Further, the source of water of most hot springs contains a relatively stable quantity of solutes which will at-least be the major source of most essential nutrients. Thus hot springs offer a unique habitat where a unique group of blue green algae flourish. However, few identical assemblages of microorganisms can be found on a global scale even when the physico-chemical environment appears to be the same. The global distribution of most thermal blue green algal species is discontinuous to a much higher degree than the patchiness and clustering of hot spring habitats themselves (Castenholz, 1969).

The species composition of only few hot spring groups of the globe has been studied (Castenholz, 1970; Castenholz and Wickstorm, 1975; Brock, 1978; Ward *et al.*, 1989). The well studied hot spring clusters are Hunter's in southern region and Yellowstone national park of USA (Castenholz, 1973). Blue green algal assemblage and their upper temperature limit was also studied in the insular thermal areas of Iceland, Newzeland, Indonesia, Phillippines, Japan and Great Britain (Castenholz, 1960; Jørgensen and Nelson, 1988, Yamaoka *et al.*,

1978; Pentecost, 1995). Few scattered reports on the blue green algal floristic composition and physico-chemical properties of thermal springs of India are also available (Gonzalves, 1947; Prasad and Srivastava, 1965; Thomas and Gonzalves, 1965; Vasishta, 1968; Jha and Kumar, 1986). The species list from hot springs so far compiled is not adequate for establishing a geographical distribution of most species. However, *Mastigocladus laminosus* was found to be cosmopolitan in the thermal springs. Two other species, *Synechococcus lividus* and *Oscillatoria terebriformis* were also common but in a restricted range. A few other species of the genera *Phormidium, Oscillateria, Calothrix, Pleurocapsa, Aphanocapsa* and *Synechococcus* are the common forms in many hot springs of the world (Castenholz, 1970). All these blue green algae are larger in size than that of bacteria of the same habitat, seldom form resting spores even at higher temperature and grew showing extreme stability towards heat of individual metabolic processes such as photosynthesis. Maximum photosynthetic efficiency of thermophilic blue green algae taken from hot springs was observed at the high temperatures of the natural habitats (Brock, 1967) showing that metabolically these organisms are specialized and are different from those of the blue green algal species thrive higher temperature and desiccation in the terrestrial habitats.

1.4 THE SCOPE OF THE BOOK

The cells of blue green algae (cyanophytes/cyanobacteria) are surrounded by mucilaginous external layers which have been named as sheath, capsule, slime or mucilage. The extracelluar investments occupy a strategic position at the outermost region of blue green algal cell envelope and there by play a fundamental role in interaction of cells with their environment (Costerton *et al.*, 1987). These external layers increase considerably in thickness in response to unfavourable environmental conditions there by providing protection to the cells under hostile environments. Like bacterial external layers, blue green algal extracelluar investments probably also provide protection against phagocytic predation, lysis by bacteria, virus and desiccation. It has been reported that capsule produced by *Microcystis* bloom in eutrophic water bodies and corresponding *Microcystis* species in culture accumulate metal ions (Parker *et al.*, 1996). External layers of blue green algae can also concentrate the essential trace elements present in submarginal concentrations in aquatic environments (Uhlinger and White, 1987). The occurrence of several functional groups (carbonyl, carboxyl, hydroxyl and sulphate) would provide various attachment sites for cations (Elchhron, 1975), and there by facilitate nutrient uptake into the cells. By concentrating nutrient at cell surface the organism could proliferate in an environment otherwise unsuitable due to an insufficient nutrient supply, and also remove excessive nutrients from an environment which otherwise would be polluting the same. The sheath of certain species bind to toxic chemicals which in excess amounts are harmful to the cells (Weckesser *et al.*, 1988). This shows that the external sheath layers might also have a protective function.

Since several blue green algal species possessing a definite external envelope layer around their cell wall were found flourishing in water bodies polluted by industrial effluents, especially those receiving liquid wastes from molasses distillery of Aska, sponge-iron factory of Palasponga and fertilizer factory of Talcher (all in the Orissa State of India), the ecophysiology of these organisms which thrive in such adverse environments has been presented. Chapter III focussed on the organisms occurring in habitats polluted by industrial effluents, their correlation with the physico-chemical characteristics of the polluted waters, their adaptive morphological features and the physiological characteristics which regulate their dominance in such habitats. In chapter IV, occurrence, growth and nitrogen fixation of blue green algae species in pesticide burdened rice field soils have been described with an aim to screen ecologically adaptive and stress compatible strains of diazotrophic blue green algae for their use as biofertilizer in rice cultivation which is of immense importance for an agriculture based country like India.

Another proposition concerning the ecological significance of slimes and sheath of blue green algae is that these external layers contribute significantly in the formation of microbial crusts/mats on the soil and subaerial habitats throughout the world from tropics to polar regions, and from temperate climates to extreme arid deserts (Starks *et al.*, 1981; Scherer *et al.*, 1984). The notable features of these microbial crusts/mats in arid regions are that they are dominated by sheath forming species of filamentous blue green algae (Campbell, 1979; Belnap and Gardner, 1993; Danin *et al.*, 1989; Lange *et al.*, 1992). On the exposed rock surface of temples and monuments of India, characteristic blackish brown crusts/mats comprising of blue green algae were observed even during mid summer months when the temperature on the rocks exceeds 60° C coupled with strong solar insolation and extreme dryness. Survival under such extreme environment is possible only if the microbial community develops mechanisms that allow it to remain viable and dormant when desiccated for longer duration. In chapter V, the morphological features and ecophysiology of a blue green alga which occur predominantly on the rock surface of several temples exposed to high temperature, direct sunlight and desiccation during summer seasons of the tropics, has been presented. Ecophysiology of blue green algal species occurring in the thermal springs of Orissa state of India has also been dealt with in chapter VI since they are the natural environments of great antiquity and relative constancy where these organisms have evolved to meet the environmental challenges of high temperature.

2

Materials and Methods

General procedures common to the experiments depicted in the different chapters are described in this chapter. In additions to these specific methods can be found in separate sections within each chapter.

2.1 CULTURE METHODS

2.1.1 Culture vessels

Corning or Borosil make glass vessels were used throughout the experimental work. Hard glass test tubes (15 x 150 mm) and conical flasks (100 ml capacity) were used for culturing blue green algae. The test tubes and conical flasks were stoppered by non-absorbant cotton wool plugs.

2.1.2 Cleaning methods

Culture vessels and other glasswares used in the experiments were first cleaned with liquid detergent solution in running tap water and dried. Then the glasswares were immersed in chromic acid (potassium dichromate in concentrated sulphuric acid, 1% w/v) overnight followed by repeated washing in running tap water and finally rinsed with double distilled water. Washed glasswares were dried in a hot air oven at 100° c for 6-8 hours and used for the experimental work.

2.1.3 Growth media

Allen and Arnon's medium (1955), Kratz and Myers medium (1955) or BG 11 medium (Rippka *et al.*, 1979) were used' for culturing blue green algae. Stock solutions of analytical grade chemicals (BDH/Qualigen/Merk/SRL) were prepared in double distilled water, sterilized and stored in refrigerater for subsequent use. To prepare 1 litre of culture medium, 1 ml from each stock was added drop wise into a clean volumetric flask containing double distilled water and the contents were stirred continuously during addition. After addition of the nutrients, the final volume of the flask was adjusted to 1 litre. The pH of the culture medium before autoclaving was always maintained at 7.5.

2.1.4 Sterilization

The culture medium, all other inorganic chemicals and the glasswares used were sterilized at 10.35 kPa pressure for 20 min in an autoclave (Yorco, India). Organic compounds when added to the culture medium were filter- sterilized using Sartorius membranae filter before addition.

2.1.5 Culture condition

Unless otherwise stated the cultures were maintained in culture racks in a temperature controlled culture room at 25±1° C under continuous light at an intensity of 7.5 W/m^2 from fluorescent tubes.

2.1.6 Inoculation

Inoculation was carried out in a bacteria free inoculation chamber, specially designed for the purpose. Before inoculation, all the glasswares and other aids necessary for inoculation were kept in the chamber and irradiated with UV-C (main output at 254 nm). Aseptic condition was always maintained during inoculation. For subculture from stocks, inoculation into fresh agar containing tubes (1% w/v agar agar in nutrient mediums was done) by using a platinum wire loop. For experiments, homogenous suspension of each organism was prepared using a sterile glass tissue homogenizer. Equal amount of the inoculum (culture suspension) was pipetted into each tube/conical flask containing sterile culture medium. The organisms at their exponential growth phase was always used as the inoculum for experiments.

2.1.7 Isolation and determination of blue green algae

Samples were collected from the natural habitat in sterilized screw-cap bottles. The samples were observed soon after in the laboratory using phase-contrast research microscope. The organisms were isolated by suspending and subsequently diluting the samples in the culture medium, and incubated at room temperature under continuous illumination at different light intensities (7.5 W/m² and 12 W/m²). After one week of enrichment, culture materials were plated on medium solidified with 1% agar (Difco Bacto). Uni-algal cultures were obtained by picking up single filaments or small colonies of uni-cellular species using sterile pipette. Attempts were made to obtain axenic culture of each of the isolated organisms by treating with low concentrations of antibiotics as per the method followed by Pattnaik (1964). For checking axenity, the organisms were incubated in bacteriological test media for 24 hours in the dark (glucose 20 g + yeast extract 5g + Bacto agar 15 g in 1 litre distilled water). Absence of milky or turbid appearance in the medium indicated absence of bacteria in the culture. This was further confirmed by observing under a phase-contrast microscope.

The cyanobacterial forms were identified following Geilter (1932), Desikachary (1959), Starmach (1966), Anand (1989) and Komarek (1993). The basis of classification of Desikachary (1959) was followed for enumeration of taxa. Where the classification of blue green algae up to species-or genus level was problematic, the system of Rippka *et al.* (1979) was used. Photographs of the organisms was taken in a Meij ML-TH-05 trinocular research microscope using Nikon F 801 camara.

2.2 EXPERIMENTAL METHODS

2.2.1 Growth measurements

All the experiments were carried out in cotton stoppered 15x150 mm culture tubes or 100 ml conical flasks containing 10 ml or 25 ml medium respectively and 1 ml of inoculum suspension. Growth was determined by measuring absorbance of the homogenized culture suspension at 760 nm in a Systronics 105 or Hitachi U-2000 UV-Vis spectrophotometer with reference to blank containing the culture medium. The objective of measuring growth of a pigmented microorganism at 760 nm is that at this wavelength absorption by pigment is minimum and the optical density here is a function of light scattering by the organism. Selecting a suitable wavelength for growth measurements was necessary because unless the rate of growth and pigment metabolism are same, optical density of a microbial suspension at pigment absorbing wavelengths will not truly reflect the growth of the organism (Venkataraman, 1969). To avoid interference by pigment absorption, some laboratories have choosen the specific wavelengths of the spectrum like 500-550 nm (Thomas and David, 1971), 750 nm (Mc Donald *et al.*, 1970; Fujita *et al.*, 1987), 760 nm (Adhikary, 1983), 800 nm (Goedheer and Kleienchammans, 1977) etc. The method of growth measurement of filamentous blue green algae by monitoring light scattering at 760 nm as described earlier (Adhikary, 1983) has been followed here.

A calibration curve was plotted for the dry weight values of the organisms versus optical density of the same cell suspension at 760 nm. It expressed a linear relationship up to a optical density value of 1.0 indicating the reliability of direct optical density measurement at 760 nm in estimating the growth of the organisms.

For certain experiments, growth was also estimated on a dry weight basis. Samples of cultures were centrifuged and dried in an oven at 100° C to constant weight. Drying for prolonged period or drying over vaccum yielded essentially the same results.

2.2.2 Statistical analysis

Triplicates were set up for each set of experiments. Mean value of 3 replicates of each experiment ± S.D. has been presented in the text.

3

Ecophysiology of Blue Green Algae in Habitats Polluted by Industrial Effluents

One of the main source of pollution of water bodies in developing countries is due to discharge of untreated industrial wastes containing various toxic substances. When the untreated industrial effluents are discharged into water bodies, they often affect the aquatic flora adversely and also degrade the water quality. Pollution caused by industrial wastes can be evaluated by physico-chemical and biological parameters of the habitat. Chemical analysis gives quantitative informations about the waste constituents, a requirement which is basic for interpreting the results as well as for the subsequent treatment of the waste. However, it is not always possible to assess the extent of pollution only by chemical data and supplementary biological data are also required. Studies on the biology of polluted habitats have revealed that the organisms living in the area provide a more sensitive and reliable measure of the condition than do pure physical and chemical determinations (Copeland, 1967; Palmer 1969, 1982). The past two decades have witnessed an increasing awareness of the important role played by algae in the biology and ecology of polluted waters. These organisms serve as good tool for assessment of the effect of waste effluents of sewage and industries because of their simple structure, preferential aquatic habitat and easy handling in pure cultures under laboratory conditions. These characteristic features enable them to adapt well in variously altered or polluted environments.

In this chapter physico-chemical characteristics of the untreated/partially treated effluents of a Molasses distillery, Sponge-iron factory and Fertilizer factory (all located in the Orissa state of India), their effect on the occurrence of different aquatic microorganisms in the surrounding habitat, and evaluation of the toxicity of industrial effluent using blue green algae as the test system has been presented.

3.1 MOLASSES DISTILLERY, ASKA

The Molasses distillery at Aska has discharged large quantities of untreated effluent to the surrounding rice fields since 1970. As farmers of this area are dependent mainly on the rice crop it creates intense fear among them on the quality of rice field soils. The effluent discharged directly from the factory was analysed for the presence of various inorganic and organic chemicals. The effluent affected areas around the distillery were survayed for seasonal occurrence of various algal forms. In addition, effect of graded concentrations of the effluent on the growth of a selected blue green alga which occurred throughout the year in the polluted site has been presented.

The main objectives were:

(i)   to find out whether the organism isolated from the polluted habitat could be made to grow in polluted water under laboratory conditions and to evaluate its nutrition status,

(ii)   to determine whether or not the effluent could support the growth of blue green algae in light and dark, either directly or after dilution with culture medium, since it has been emphasized that waste organic matters serve as the main nutrient influencing the heterotrophic growth of certain algae in natural environments (Zajic and Chiu, 1970), and

(iii)   possible use of blue green algae capable of utilizing distillery effluent as biofertilizer for rice crop.

3.1.1 The study site

The study site is located at Aska (19° 16′ N, 84° 55′ E), 170 Km away from Bhubaneswar surrounding the Aska Co-operative Sugar Industries (Fig 1).

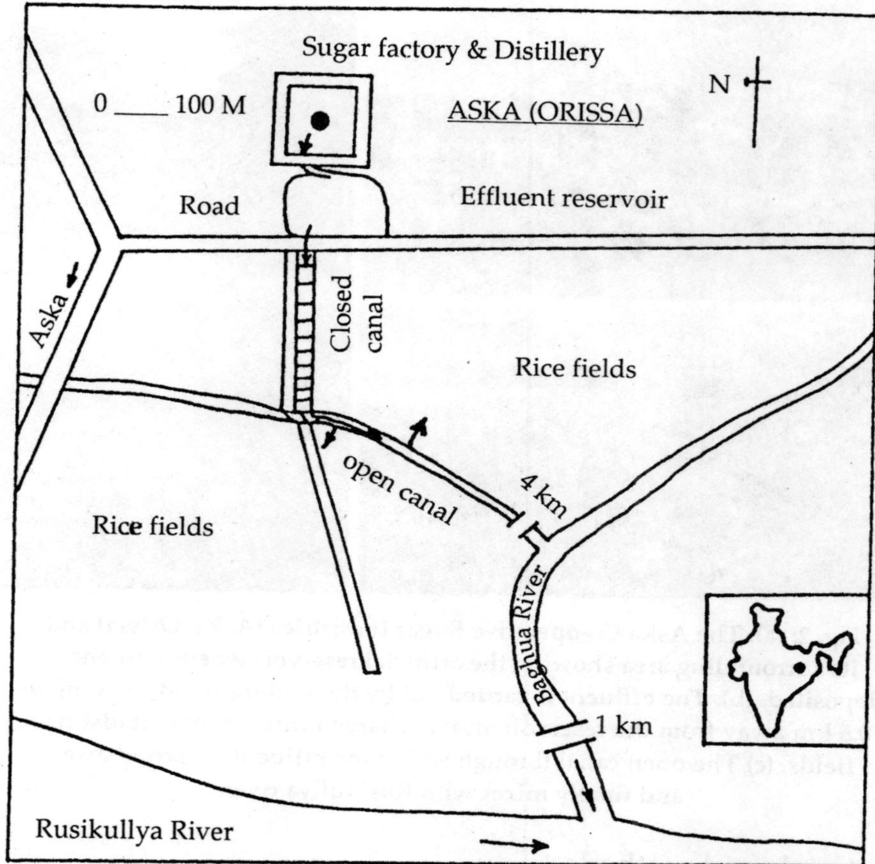

Fig.1: Study site showing location of the sugar factory and distillery (Aska Co-operative Sugar Co., Aska) and the passage of waste water from the discharge point of the distillery to Rusikullya river.

The factory discharges untreated effluent to the surrounding environment from September to February each year. About 3 to 4 million gallons of effluent are discharged from the factory per day. The effluent is stored in an artificial reservoir adjacent to the factory (Fig. 2a) and then carried by an underground canal to a small rivulet (open canal) at 0.5 Km away from the discharge point (Fig. 2b) which ultimately terminate into Rusikullya river (Fig. 1, Fig. 2c). The study area around the effluent canal received much damage due to the waste water discharged from the factory. In the underground canal there are some leakages through which the effluent leaked out to the adjoining rice fields. In rainy season, the effluent stock and the open canals are often overflooded with water. The effluent enters the surrounding rice fields covering about 10 square kilometer area and pollute the environment.

Fig. 2: (a). The Aska Co-operative Sugar Industries (Aska, Orissa) and
its surrounding area showing the artificial reservoir where effluent is
deposited. (b). The effluent is carried out by the underground canal up to
0.5 km away from the reservoir and discharged into a canal amidst rice
fields. (c) The open canal through which the effluent is carried away
and finally mixes with Rusikullya river.

3.1.2 Materials and methods

Effluent sample were collected in narrow mouth glass and polythene bottles.
The pH of the effluent was recorded on the spot at the time of collection. The
methods followed for analysis of the effluent was based on APHA (1971, 1989).

The physico-chemical characteristics of the effluent (collected from the
discharge point of the factory) is given in Table-4.

Table 4 : Physico-chemical characteristics of the distillery effluent (Aska,
Ganjam Dist, Orissa, India). (Mean ± S.D. values of 12 observations, all
values except pH and reducing sugar content are in mg/litre)

Characteristics	Mean ± S.D. values	Tolerance limit for industrial effluents discharged on land for irrigation purpose. IS: 3307 (1965)
Colour	Reddish brown	
Odour	Smell of burnt sugar	

Characteristics	Mean ± S.D. values	Tolerance limit for industrial effluents discharged on land for irrigation purpose. IS: 3307 (1965)
Total solids	5225.0±3.08	2100 max
PH	4.8±0.3	5.5-9.0
Dissolved oxygen	0.5±0.1	
Biochemical oxygen demand (B.O.D.) for 5 days at 20° C	3666.0±3.28	500 max
Chemical oxygen demand (C.O.D.)	2321.5±5.75	
Total nitrogen	86.7±2.23	
Sulphate	208.6±1.4	1000 max
Chloride	745.2±2.55	600 max
Phosphate	9.3±0.28	
Sodium	25.0±0.72	60 max
Calcium	68.2±0.66	
Iron	5.6±0.13	
Magnesium	16.4±0.18	
Potassium	89.3±3.15	
Reducing sugar[a]	4.52-6.0%	

a) The effluent was first hydrolyzed to break down the polysaccharides and disaccharides and analyzed. The results given is the amount of total reducing sugars present in the effluent.

The effluent was highly acidic and contained objectionable amounts of total dissolved solids, chlorides and the biochemical oxygen demand value was very high. Hence they do not meet the ISI standards (IS: 3307, 1965). The effluent, rich in various algal nutrients such as nitrogen, phosphate and calcium, iron and potassium salts and high amount of reducing sugars (4.5-6.0%), supported the growth of 27 species of algae in the polluted area, of which the members of blue green algae were dominant (Table-5).

Table 5 : Seasonal variation and occurrence of different organisms in the distillery effluent polluted habitat

Organism	Summer Feb.-May	Rainy June-Sept.	Winter Oct.-Jan.
Chroococcus minutus (KÜTZ.) NÄG.	++	+	+
Gloeocapsa atrata (TURP.) KÜTZ.	+	+	+

Organism	Summer Feb.-May	Rainy June-Sept.	Winter Oct.-Jan.
Merismopedia glauca (HER.) NÄG.	+	-	-
Oscillatoria formosa BORY.	-	+	+
Oscillatoria redekei GOOR	++	+	+
Oscillatoria agardhii GOM.	-	-	+
Phormidium ambiguum GOM.	++	+	+
Phormidium molle (KÜTZ.) GRUN.	+	-	-
Phormidium purpurascens (KÜTZ.) GOM.	+	-	+
Spirulina major KÜTZ.	+	-	+
Anabaena sp.	++	+	++
Anabaena constricta ZAFAR	+	-	++
Anabaena fertilissima RAO	-	+	+
Cylindrospermum muscicola KÜTZ.	-	+	+
Cylindrospermum stagnale KÜTZ.	-	-	+
Calothrix marchica LEMM. var. *intermedia* RAO	+++	++	+++
Scytonema simplex BHARDW.	-	+	+
Scytonema schmidlei DE TONI	++	+	++
Westiellopsis prolifica JANET	-	+	+
Pediastrum boryanum (TURP.) MENEGH.	+	-	+
Pediastrum duplex MEYEN.	+	-	+
Scenedesmus quadricauda (TURP.) BREB.	++	+	+
Peridinium cerasus PAULSEN	+	-	+
Perinidinium cerasus PAULSEN	+	-	+
Peridinium depressum BAIL.	-	-	+
Navicula hendeyii SMITH	+		+
Nitzschia acicularis SMITH	++	-	+
Nitzschia seriata CLEVE	-	-	+

+, occurred in less than 20% of the samples; ++, occurred in approximately 50% of the samples; +++, occurred in almost all the samples examined; -, absent

The heterocystous filamentous blue green alga *Calothirx marchica* Lemm. var. *intermedia* Rao was found to occur throughout the year in the study site. The alga growing in the polluted habitat showed thick sheath layers around its trichome. However, when isolated in unialgal culture by repeated sub-culturing in different conventional culture media (Fogg, 1949, Allen and Arnon, 1955, BG 11 medium of Rippka *et al.*, 1979, Kratz and Myers, 1955) filaments with thin sheath layer and a distinct basal heterocyst were produced. Kratz and Myer's medium (1955) was found suitable for the growth of the organism. Pure culture of the blue green alga was obtained following the method as described in Chapter-2. To understand why *Calothrix marchica* grew abundantly in the polluted habitat, growth of the organism in graded levels of the distillery effluent was studied. Attempts were made to decolorize the effluent by activated charcoal before using in the experiments, but even on repeated trials colourless effluent could not be obtained. Therefore, sterilized effluent, after filtration, was directly used for the experiments.

Chemical analysis showed that the effluent contained 4.5-6.0% reducing sugars; the sugars were composed mostly of glucose and fructose. The neutral sugars, liberated by 0.5 N H_2SO_4 (100° C, 4h) were identified by thin layer chromatography (solvent-ethyl acetate: pyridine: water = 12:5:4, v/v/v; staining - anilinium pthalate). Since it has been emphasized that waste organic matter serve as the main nutrient influencing heterotrophic growth of certain algae in natural environments (Zajic and Chiu 1970, Fogg *et al.* 1973), few experiments were conducted in light, in light in presence of 10^{-5} M DCMU and in the dark with distillery effluent and various organic carbon compounds as possible energy source for heterotrophic growth of the organism. The carbon compounds, mainly glucose, galactose, fructose, mannose, xylose, ribose, sucrose, sodium acetate, fructose 1,6 diphosphate, pyruvic acid, citric acid and succinic acid were autoclaved separately from the mineral salts and subsequently added aseptically to the culture media. All media were brought to a pH of 7.5 prior to sterilization. The pH of the culture media were tested from time to time during the experimental period and adjusted within the above range by aseptic addition of few drops of N/10 NaOH or N/10 HCl. Heterotrophic bacterial contamination was monitored by plating the blue green algae suspension on Difco nutrient agar supplemented with glucose. Experiments were conducted in cotton stoppered 100 ml capacity conical flasks containing 25 ml of culture medium. Pure inorganic nitrogen free medium was always kept as control. All the experiments were done in triplicate and the inocula (equivalent to 1.0 mg dry weight each) were taken from exponentially growing cultures. The experiments were carried out in a culture room at 25±1 ° C and at 7.5 W/m² light intensity or in the dark. Growth of the blue green alga was determined on dry weight basis.

3.1.3 Results and discussion

3.1.3.1 Effect of graded concentrations of the effluent on the growth of *Calothrix marchica*

The blue green algae did not grow in the pure waste waster of the distillery (Fig.3).

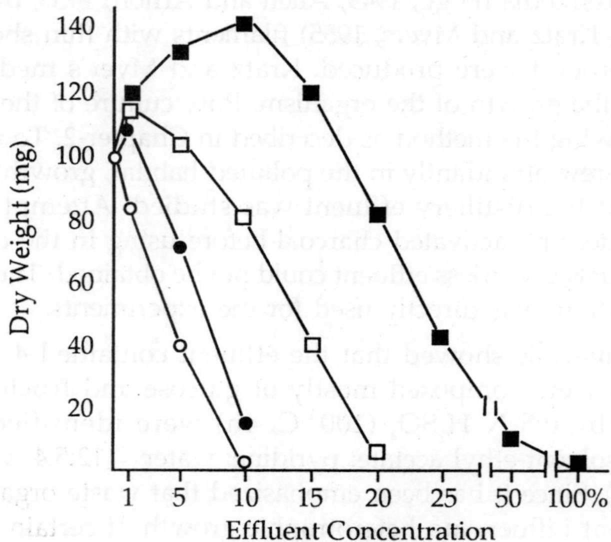

Fig.3: Growth (dry weight basis) of *Calothrix marchica* in different concentrations of the effluent. Values represent per cent increase or decrease over control (basal inorganic medium).
◯, different concentrations of the effluent prepared in glass distilled water;
●, different concentrations of the effluent prepared in glass distilled water and the pH adjusted to 7.0; ▢, different concentrations of the effluent prepared in basal inorganic medium;
■, different concentrations of the effluent prepared in basal inorganic medium and the pH was adjusted to 7.0.

This high degree of toxicity may be due to the acidic pH of the effluent and/or the excessive nutrient status of the pure effluent. A dilution of the medium with glass distilled water did not support the growth of the organism. When the waste water of the distillery was diluted with basal inorganic medium, the effluent only at low concentrations supported the growth of *C. marchica*. The growth was increased at low concentrations of the effluent when the latter was diluted with the culture medium over that when the effluent was diluted with distilled water, though the effluent at higher concentrations, whether diluted with basal inorganic medium or distilled water, decreased the growth rate of the organism. It is known that dilution of an acid solution raises the pH to some extent without further adjustment with alkali. So the better growth of *C. marchica* in the culture containing low concentrations of the effluent diluted with basal inorganic media may be due to the increase in pH level and additional nutrients. When the waste water was diluted with inorganic medium after the pH was adjusted to pH 7.0, the blue green alga could grow at an enhanced rate over the control values at lower dilutions (1-10%, v/v) in light (Fig. 3) and upto 50% (v/v) dilution in darkness (data not given); highest increase in growth of the experimental organism was obtained in the cultures containing 10% (v/v) of the effluent.

3.1.3.2 Effect of exogenous substrates on the growth of *Calothrix marchica* in mixo-, photo- and chemo-heterotrophic culture

The various exogenous carbon compounds, the range of concentrations used, and the resulting growth in light, in light in presence of 10^{-5} M DCMU, and in the dark are summarized in Table-6. The exogenous substrates such as glucose, fructose, galactose and sucrose, each supported the growth in light in presence or absence of DCMU and in the dark; in each case 0.5% concentration was optimum. Sucrose and lower concentration of neutralized distillery effluent were found to be the best to support the heterotrophic growth of the organism. This was also confirmed by long term experiments on the effects of sucrose (0.2, 0.5 and 1.0%, w/v) and distillery effluent (10%, v/v) on the growth of *C. marchica* in light and dark (Fig. 4). The alga did not grow in the pure basal inorganic medium in the dark or in light when the CO_2 assimilation was indirectly blocked with the herbicide DCMU (Rippka 1972). But when the inorganic medium was supplemented with different concentrations of sucrose or distillery effluent, growth of *C. marchica* increased to a significant extent (Fig. 4). This mode of growth increase in culture density was totally dependent on exogenous organic compounds. Light was highly stimulatory for the heterotrophic growth of *Calothrix*. Light could immediately and efficiently enhance the growth of the organism in the culture supplemented with exogenous substrates. The photo-stimulation and assimilation of organic compounds may be connected with the abundant supply of high energy phosphate generated in the photochemical reaction which could support the energy requiring transport of exogenous substrates (Stanier, 1973; Smith, 1982) and their subsequent assimilation in the cells.

Table 6 : Growth of *Calothrix marchica* on various substrate concentrations and distillery effluent in light, in light in presence of 10^{-5} *M* DCMU and in the dark. Cultures were incubated for 15 days. Results represent the substrate induced growth over the control (cultured in basal inorganic medium without addition of exogenous substrate). (++, dense growth; +, fair growth; t, trace of growth; -, no growth)

Compound	Concentration (%)	Growth condition		
		Light	Light+10^{-5}M DCMU	Dark
Glucose	0.2, 0.5	+	t	t
	1.0	t	t	t
Fructose	0.2, 0.5	+	+	+
	1.0	+	t	t
Mannose	0.2, 0.5, 1.0	-	-	-
Galactose	0.2, 0.5	+	t	t
	1.0	t	t	t

Compound	Concentration (%)	Growth condition		
		Light	Light+10^{-5}M DCMU	Dark
Xylose	0.2, 0.5, 1.0	-	-	-
Ribose	0.2, 0.5, 1.0	-	-	-
Sucrose	0.2	+	t	t
	0.5	++	+	+
	1.0	+	t	t
Sodium acetate	0.2, 0.5	t	t	t
	1.0	-	-	t
Fructose 1,6 diphosphate	0.2, 0.5, 1.0	-	-	-
Pyruvic acid	0.2, 0.5, 1.0	-	-	-
Citric acid	0.2, 0.5, 1.0	-	-	-
Succinic acid	0.2, 0.5, 1.0	-	-	-
Distillery effluent	1.0, 5.0	+	t	t
	10.0	++	+	+
	15.0, 20.0	t	t	t
	50.0	-	-	-

Microscopic examination revealed that a new type of filaments with certain specific features of *Calothrix marchica* appeared in the mixotrophic and chemoheterotrophic culture. In the new type of filaments the trichomes were elongated and surrounded by thick sheath layers. Sheath formation was not observed in the medium supplemented with mannose, pentoses, acetate or organic acids which did not support photo-and chemoheterotrophic growth of the organism.

In the mineral medium in light, the filaments were straight, at the base 8.6±0.26 µm broad with a close thin and colourless sheath without terminal hair; the cells broader than long and basal heterocysts hemispherical (Fig. 5 a). In mixotrophic culture, in presence of fructose or sucrose in light, a thick sheath layer was developed around the trichome; at the base the filaments were 11.8±0.35 µm broad of which the sheath layer around the cells was 3.5±0.12 µm in thickness (Fig.5 b). But in sucrose or fructose supplemented culture in dark, and in light in the presence of DCMU, the giant filaments of the cyanobacterium were produced in which the trichomes devoid of basal heterocyst were much elongated with cells which were longer than broad and enveloped by a distinct sheath layer; the filaments were 9.8±0.22 µm broad of which the sheath layer around the cells was 2.4±0.31 µ m in thickness (Fig. 5 c). Production of giant forms in the blue green

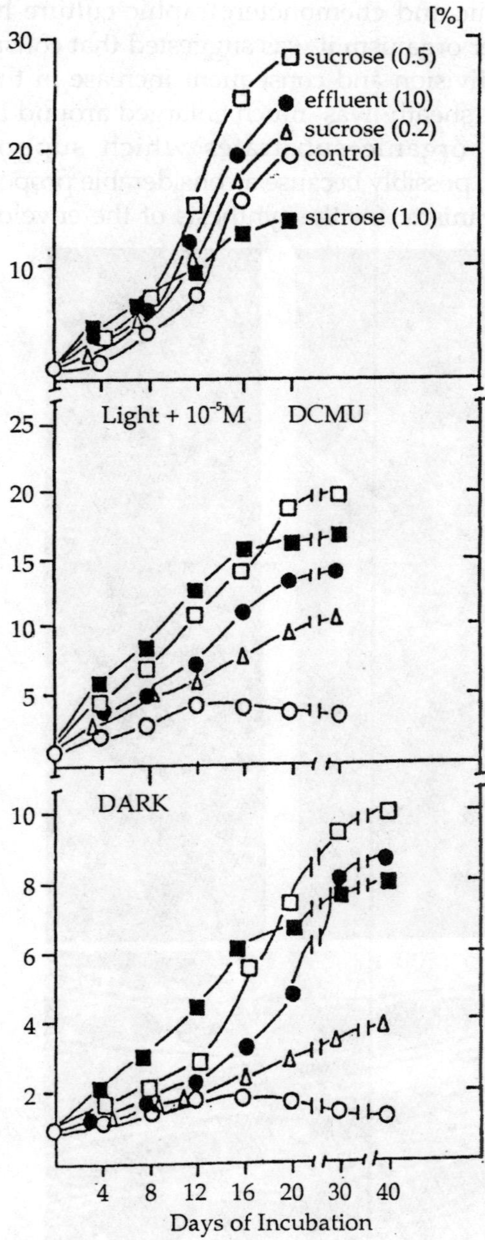

Fig.4: Growth of *Calothrix marchica* in presence or absence of sucrose (0.2, 0.5, 1.0%, w/v) and distillery effluent (10%, v/v) in light, in light + 10^{-5} M DCMU and in the dark. ○,control; ●, distillery effluent (10%); △, sucrose (0.2%); □, sucrose (0.5%); ■, sucrose (1.0%)

alga *Westiellopsis prolifica* (Adhikary, 1985) and a green alga *Chlorella* (Rodriguez-Lopez, 1965) in mixotrophic and chemoheterotraphic culture has been reported earlier, of which in the latter organism it was suggested that consumption of sugars leads to inhibition of cell division and consequent increase in the dry weight and gigantism of the cells. The sheath· was much enlarged around the trichome of *C. marchica* in presence of organic substrates which supported mixo- and chemoheterotrophic culture, possibly because a considerable proportion of the sugars was assimilated by the organism for the synthesis of the envelope layer.

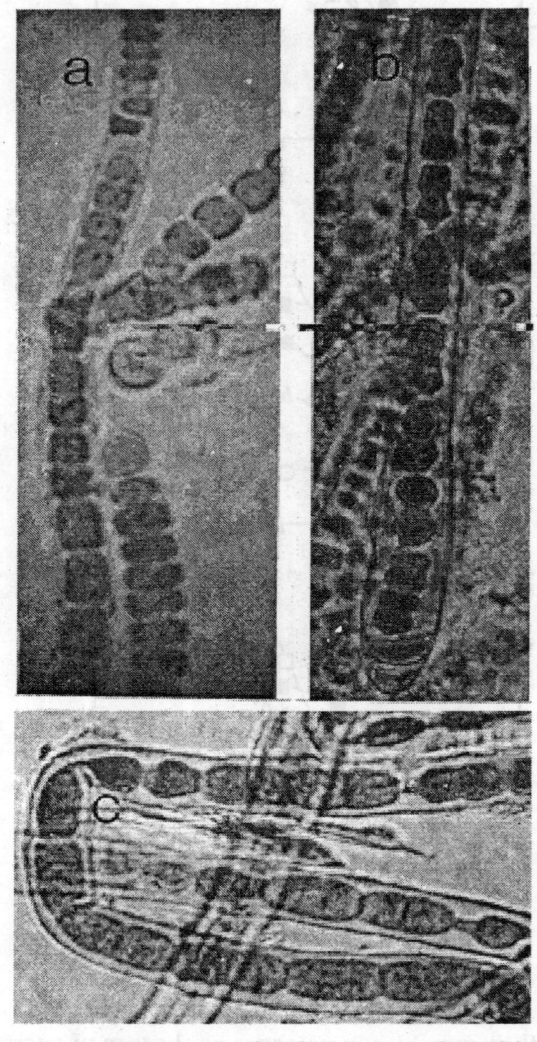

Fig. 5a-c: Light microscopic photograph of *Calothrix marchica*, (a) from basal inorganic medium in light (x445); (b) from fructose supplemented medium in light showing thick sheath layer around the trichome (x385); (c) from fructose supplemented medium in dark showing elongated trichome surrounded by thick sheath layer (x 360).

In the present case the abundant growth of *C. marchica* throughout the year in the effluent receiving area and relative absence of other organisms indicate the potentiality of *Calothrix* to grow in such adverse conditions which are not conducive to growth of other blue green algae. Possibly most of the organisms occurring in the effluent affected area could not tolerate acidic pH and could not utilize the waste organic matter for their growth and development. *Calothrix marchica* tolerated high concentrations of the distillery waste as compared to other blue green algae (Chapter 3.5) and also produced new type of filaments where the trichomes were elongated and surrounded by sheath layers in distillery effluent supplemented mixo- and chemoheterotrophic cultures. Since the formation of sheath was not observed in the medium with various sugars and organic acids which did not support heterotrophic growth, it is possible that the overall response of *Calothrix marchica* to distillery effluent is determined by an individual adaptive response. The predominance of a specific type of response is determined in turn by physiological characteristics of the species. The flexibility of its anabolic and energy metabolism built into the genetic apparatus of the blue green algae is possibly realized in presence of distillery effluent containing higher amounts of organic wastes. This factor might lead to the appearance of a new structural type of *Calothrix marchica* with elongated trichomes enveloped by thick sheath layers corresponding to the reorganized metabolic apparatus of the organism in response to the new environmental conditions. These characteristics are perhaps responsible for enabling *Calothrix marchica* to grow in the distillery effluent polluted habitat.

3.2 EFFECT OF DISTILLERY EFFLUENT AND HETEROTROPHICALLY GROWING NITROGEN-FIXING BLUE GREEN ALGAE ON THE GROWTH AND YIELD OF RICE

The problem of disposal of liquid wastes from distilleries has presented one of the most urgent and serious environmental crisis in India and in other developing countries. The rice fields are being affected by the effluent from nearby distilleries. A large number of heterocystous and non-heterocystous forms of blue green algae have been reported in the distillery effluent polluted rice-fields near Aska. The nitrogen-fixing blue green alga *Calothrix marchica* Lemm which grows luxuriantly in the rice fields affected by distillery effluent was reported to grow both mixotrophically and heterotrophically in the dark by utilizing low concentrations of the neutralized effluent. The effluent from distilleries contains high concentrations of nitrogen, phosphorus and potassium salts. The nitrogen-fixing blue green algae and the effluent might be interacting with each other in the natural environment affecting the growth and yield of rice.

Nitrogen-fixing blue green algae are at present used as bio-fertilizer in tropical and sub-tropical countries (Agarwal, 1979). The use of nitrogen-fixing blue green algae as an alternative nitrogen source is economic compared to inorganic nitrogen fertilizer (Venkataraman, 1972). Based on the natural ecology of these organisms, a simple rural oriented open-air method has been developed (Vankataraman, 1977). But since economy is the most important consideration in any application, nitrogen-fixing blue green algae capable of growing at an enhanced rate by utilizing organic wastes is of considerable interest where large quantities of materials can be produced for their use as bio-fertilizer within a short time.

A number of heterocystous blue green algae namely, *Anabaena* sp., *Calothrix marchica* and *Scytonema schmidlei* which occurred in the distillery effluent polluted rice-fields have been isolated. An attempt was made for open-air cultivation of nitrogen-fixing blue green algae in bulk, singly and combination of different strains, by utilizing low concentrations of distillery effluent. Further, the effect of various combinations of distillery effluent and the alga *Calothrix marchica* Lemm. var *inter media* Rao on the growth and yield of an early variety of rice MTU-17 was studied.

3.2.1 Materials and methods

Five grams of sundried *Calothrix marchica* and same quantity of mixture of heterocystous blue green algae *Anabaena* sp., *Calothrix marchica* and *Scytonema schmidlei* were used as starter culture. Experiments were carried out in 10 litre capacity wide mouth glass jars with or without different concentrations of commercial NPK (5, 25 and 50 g) and distillery effluent (2%, v/v) prepared in Kratz and Myer's medium (1955). The concentration of the effluent >2% in the glass jars did not allow light to penetrate till the bottom of the vessels. Thus, 2% (v/v) of the effluent was used for mass cultivation of blue green algae. The contents were sterilized and then the organisms were inoculated and grown under direct sunlight for 15 days. The blue green algal materials of the glass jars were filtered through a bolting silk cloth, transferred to pre weighed whatman no 1 filter paper, dried in a hot air oven at 100° C for 24 hours and weighed. Total nitrogen content of the cultures of the organism was determined following Herbert *et al.* (1971). The amount of nitrogen fixed per culture was calculated after deducting the amount of nitrogen present in the inoculum and in the culture medium at the begining of the experiment.

Pure line seeds of the rice variety MTU-17 were obtained from the Orissa University of Agriculture and Technology field experimental station, Ratanpur, Ganjam. The seeds were soaked for 12 h in distilled water at 28±2% C and placed in sterile petri dishes. Fifty seeds were uniformly spaced on filter papers and were treated with water and the test solution (different concentrations of the distillery effluent) as and when necessary. The petri dishes were illuminated from above by fluorescent tube lights to give an intensity of 2,400 lux at seed level. The number of seeds germinated at each 12 h interval was counted and the percentage of germination at 96 h of treatment was presented here. Germination refers to the initial appearance of the radicle by visual observations. Number of seeds germinated was expressed on a percent basis.

A second set of experiments was conducted in petri dishes containing vermiculite. Seeds were germinated in double distilled water and seven-day-old seedlings were transferred to the petri dishes. Twenty healthy seedlings were uniformly spaced in each petri plate. The alga *Calothrix marchica* Lemm, isolated from the distillery effluent polluted rice-fields, was used as the biofertilizer. The organism was cultivated in 1000 ml conical flasks in nitrogen-free basal inorganic medium (Kratz and Myers, 1955). The blue green algal suspension contained 100 mg dry weight cells/litre, when used in the experiments. The cells were suspended in water, or in a filtrate of the culture medium containing substances released by live or lysing cells. The filtrate contained the nutrient salts of the growth medium. The rice plants were subjected to the following conditions: control (Hoaglnd's solution) (Hewitt, 1963); 10%(v/v) neutralized distillery effluent prepared in Hoagland's solution; blue green algae suspension (equivalent to 100 mg dry weight/ litre of culture medium); extracelluar products liberated by 100 mg dry weight of *Calothrix marchica* in the nitrogen free culture medium, and the culture medium (Kratz and Myers, 1955), which was unavoidably added to the seedlings in other treatments. Application of these treatments singly or in various combinations was made either at the start of the experiments only or additionally at regular intervals during the growth period of plants up to 20 days. Each treatment was replicated five times. The root and shoot length of each plant was measured. Dry weight determinations were made following the method described earlier.

The third set of experiments were conducted in earthen pots of 6 kg capacity, each containing 5 kg of soil. The pots along with soil were autoclaved at 15 lb pressure for 15 minutes. Twenty-day-old plants were transplanted into each pot. The following treatments were used: (1) blue green algal suspension (A), (2) 10% (v/v) neutralized distillery effluent prepared in tap water (E), and (3) blue green algal suspension and 10% neutralized distillery effluent in combination (A+E). The production of tillers, leaves and height of the plant were recorded at the age of 40, 60, 80, 100 and 110 days. The effect of various treatments on the components of yield, viz., the number of panicles emerged from 65th day till the last stage of growth, length of panicle from the last node of the shoot up to the top most grain, total number of spikelets per panicle, total number of grains per panicle, total number of empty spikelets per panicle, percentage of grains set per panicle and weight of 1000 grains in grams were compared on the 110th day of the rice plants. Grain yield per plant for each treatment was determined. The surface area of the soil in the pots was determined and from this the approximate grain yield of the rice variety MTU-17 per acre area was calculated.

3.2.2 Result and discussion

3.2.2.1 Mass cultivation of blue green algae utilizing distillery effluent

In presence of lower concentrations of distillery effluent growth as well as nitrogen fixation of *C. marchica* increased significantly over the control values (Table-

7). This indicated the beneficial effect of cultivating nitrogen-fixing blue green algae in bulk, singly or in mixed cultures, using diluted distillery effluent for their use as biofertilizer. A range of doses of commercial NPK fertilizer (40:30:30) were tested to find out which one support the growth of blue green algae in the presence or absence of distillery effluent. Yield of blue green algal biomass over the control values was increased when NPK was supplemented up to 5 g per 10 litre medium in the glass vessel in autotrophic culture. But since the effluent contained higher amount of nitrogen, phosphorus and potassium, commercial NPK at higher concentration (more than 1 g/vessel) together with the distillery effluent decreased the yield of blue green algae. It was observed that lower concentration of neutralized distillery effluent (1%, v/v) together with 1 g of NPK fertilizer per 10 litre medium supported maximum yield of nitrogen-fixing blue green algae. These results indicate the beneficial effect of use of distillery wastes for cultivating blue green algae, capable of growing mixo-tropically and fixing nitrogen in bulk for use as biofertilizer.

Table 7 : Open air cultivation of blue green algae in large scale by utilizing distillery effluent. The values represent mean ± S.D. Final yield of blue green algae is given in grams. Nitrogen fixation of *Calothrix marchica* in laboratory culture under various growth conditions is also presented.

Growth condition	Strater culture (5 g of sun dried blue green algae material)		Total nitrogen (mg/25 ml culture) of *Calothrix marchica*[*]
	Mixture of *Anabaena* sp., *Calothrix marchica* and *Scytonema schmidlei*	*Calothrix marchica*	
A. Autotrophic (in mineral medium in light)	59.1±1.85	36.8±1.90	1.7±0.26
B. Mixotrophic (2% effluent in the medium in light)	141.8±2.24	86.4±3.74	2.8±0.31
C. Autotrophic (A)			
A+NPK 1g/vessel	62.7±1.15	41.5±2.11	2.3±0.34
A+NPK 5g/vessel	74.4±2.83	54.9±3.54	1.1±0.12
A+NPK 10g/vessel	47.3±1.54	31.2±2.58	0.4±0.07
D. Mixotrophic (B)			
B+NPK 1g/vessel	145.3±1.82	98.6±4.45	3.1±0.45
B+NPK 5g/vessel	71.8±2.79	42.8±3.19	1.2±0.21

[*]   Nitrogen fixation by *Calothrix marchica* was estimated on the 15th day of growth in laboratory culture.

3.2.2.2 Effect of distillery effluent on germination of rice seeds

Germination of the rice variety MTU-17 was not affected in the presence of 1-5% (v/v) neutralized and unneutralized effluent. The inhibitory effect of effluent on germination increased with the increase in the concentration of the distillery effluent. At 50% of the unneutralized effluent, only 8% rice seeds germinated. However, when the effluent was neutralized, about 32% of the seeds germinated at 50% of the neutralized effluent. None of the seeds germinated when treated with pure distillery effluent (Table-8).

Table 8 : Effect of distillery effluent on the germination of the rice variety MTU-17. Values represent mean ± S.D.

Effluent concentration(%)	Germination (%)	
	Unneutralized effluent	Neutralized effluent
1	100	100
5	95 ± 2	100
10	67 ± 3	92 ± 3
25	44 ± 2	74 ± 2
50	8 ± 1	32 ±2
75	0	4 ±1
100	0	0

3.2.2.3 Combined effect of the distillery effluent and the blue green alga *Calothrix marchica* on the growth of rice seedlings

It has been shown previously that the neutralized distillery effluent at 10% (v/v) concentrations supported the growth and increased nitrogen-fixing capacity of *Calothrix marchica* Lemm. in light and dark. In order to find out the extent to which distillery wastes and *C. marchica* support the growth and development of the MTU-17 rice, certain experiments were carried out and the results are shown in Table-9.

Table 9 : Effect of neutralized distillery effluent (10%, v/v), nitrogen-fixing blue green algae and their extracellular products on the root length, shoot length, moisture uptake, dry weight and number of leaves of the rice variety MTU-17. Plants were grown for 20 days at 28±2% C under 2400 lux intensity. Values represent mean ± S.D.

Treatment	Root length (cm)	Shoot length (cm)	Root length/ shoot length	Moisture uptake (mg)	Dry weight (mg)	Number of leaves
Control	83±0.3	10.5±0.4	0.79	84.4±0.4	35.6±0.3	5
10% distillery effluent; pH, 7.0	7.6±0.4	13.1±0.5	0.58	93.6±0.4	38.9±0.3	6

Treatment	Root length (cm)	Shoot length (cm)	Root length/ shoot length	Moisture uptake (mg)	Dry weight (mg)	Number of leaves
Nitrogen fixing blue green algae	8.9±0.5	11.2±0.2	0.79	83.6±0.2	36.4±0.2	6
Extracellular products of blue green algae	9.1±0.2	10.9±0.2	0.83	84.7±0.3	36.9±0.3	6
The culture medium	8.1±0.3	9.8±0.6	0.82	85.1±0.2	36.1±0.1	5
10% distillery effluent; pH, 7.0 + blue green algae	11.2±0.5	14.4±0.4	0.77	95.2±0.3	42.8±0.4	7

When the blue green alga or its extracelluar products were supplied, the vegetative growth of the rice plants increased compared to control plant. The extracelluar products of the alga contained various nutrients of the culture medium and these nutrients were unavoidably added to the seedlings in the extracelluar product treatment. When rice was treated with the culture medium + Hoagland's solution, the growth and development of rice plants was comparable to that of control plants. Thus the increase of growth of rice was due to the presence of the blue green alga and its extracelluar products and not due to the addition of the culture medium. Rice plants had a greater dry weight increase in distillery effluent (10%, v/v) treatment than in presence of blue green algae and its extracelluar products. However, the root length/shoot length ratio of rice plants was low in the presence of the effluent compared to the control plants. When the distillery effluent (10%, v/v was supplemented with blue green algae, there was a significant enhancement of the growth and development of rice seedlings (Table-9). An appreciable increase in root length, moisture uptake, dry weight yield and number of leaves per plant was observed.

3.2.2.4 Combined effect of distillery effluent and *Calothrix marchica* on the growth and yield of rice

The data in Fig.6 show the average height, number of green leaves and number of tillers produced by the rice plants in presence or absence of the *Calothrix marchica* and distillery effluent (10%, v/v; pH 7.0) at different times of growth up to 110th day. The plants grew tallest in the presence of the distillery effluent. However, when blue green algae and distillery effluent were added together in the experimental pots, a relatively higher number of leaves and tillers were produced in comparison to only effluent or algal application (Fig.6). The effect of *Calothrix marchica* and distillery effluent on the following components such as number of panicles per plant, length of panicle, number of spikelets per panicle and number of grains per panicle is shown in Fig.7. The highest yield of these components occurred when the effluent and blue green algae were added together. The yield parameters like per cent of grains set per panicle, grain yield per plant, yield of rice per acre (calculated value) were much reduced for rice grown in distillery effluent only (Fig.7). The highest grain yield in rice occurred in the presence of blue green algae and effluent. The number of empty spikelets produced per panicle was the highest and the weight of 1000 grains was the lowest in the presence of the distillery effluent in comparison to the other treatments (Fig.7).

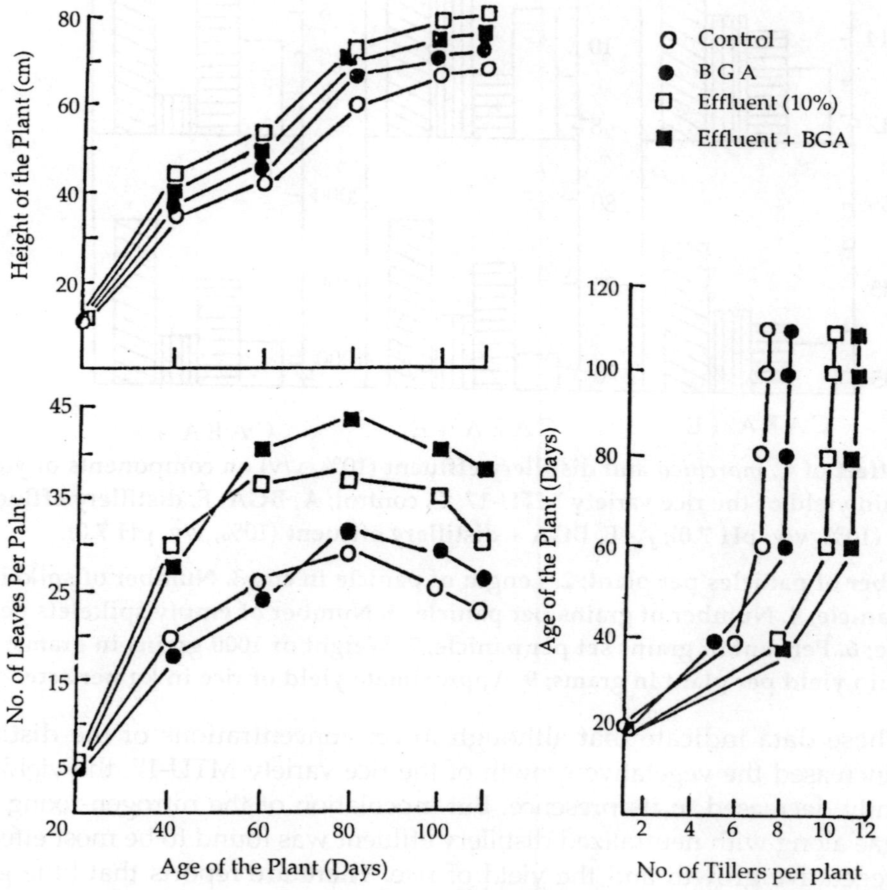

Fig.6: The effect of *C. marchica* and distillery effluent (10%, v/v) on the height, number of leaves and number of tillers of the rice variety MTU-17 at different stages of growth.

BGA = The blue green alga *Calothrix marchica* Lemm.

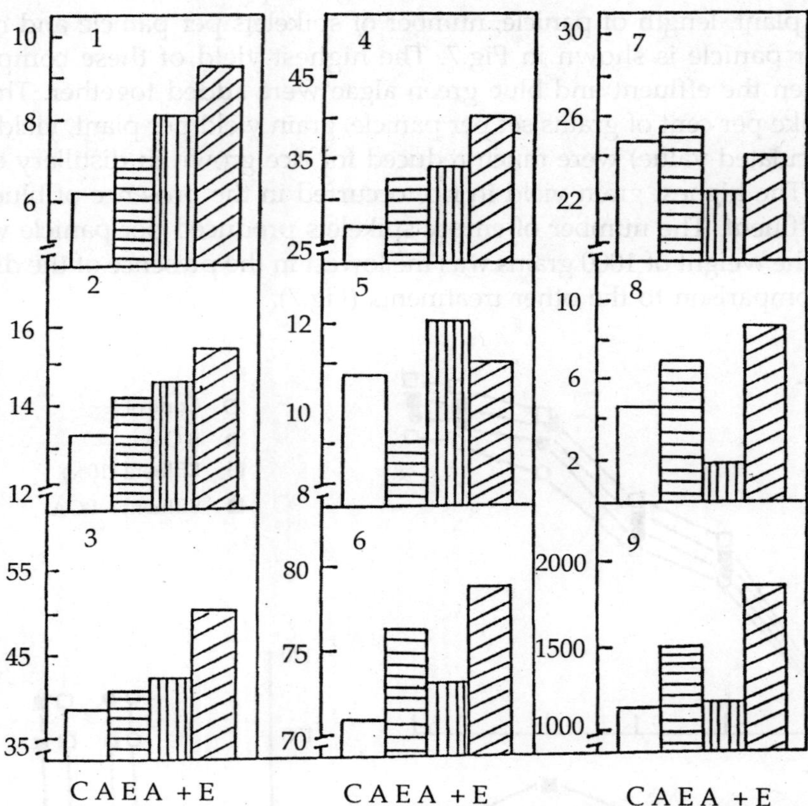

Fig.7: Effect of *C. marchica* and distillery effluent (10%, v/v) on components of yield and grain yield of the rice variety MTU-17. C, control; A, BGA; E, distillery effluent (10%, v/v; pH 7.0); A+E, BGA + distillery effluent (10%, v/v, pH 7.0).

1. Number of panicles per plant; 2. Length of panicle in cm; 3. Number of spikelets per panicle; 4. Number of grains per panicle; 5. Number of empty spikelets per panicle; 6. Per cent of grains set per panicle; 7. Weight of 1000 grains in grams; 8. Grain yield per plant in grams; 9. Approximate yield of rice in kg per acre.

These data indicate that although lower concentrations of the distillery effluent increased the vegetative growth of the rice variety MTU-17, the yield was significantly decreased in its presence. But inoculation of the nitrogen-fixing blue green algae along with neutralized distillery effluent was found to be most effective for the vegetative growth and the yield of rice. There are reports that blue green algae release biologically active compounds into the soil and that these compounds are then assimilated by higher plants, significantly enhancing their growth (Dadhich *et al.*, 1969; Venkataraman and Neelakanthan, 1967). Low concentrations of the distillery effluent increased the growth of the alga *Calothrix marchica* (Adhikary,

1987) and also the growth of rice plants. Since the nitrogen- fixing blue green algae are capable of utilizing organic wastes and low concentrations of the neutralized distillery wastes, these organisms can grow and secrete biologically active compounds which promote better growth and yield of rice. In the distillery effluent polluted rice fields, the effluent concentrations were usually too high due to low water supply. Thus, the blue green algae were not able to thrive because of the high degree of toxicity of the raw effluent and consequently do not increase the growth and yield of the rice crop.

3.3 SPONGE-IRON FACTORY, PALASPONGA

No report is available on the evaluation of the Sponge iron factory liquid wastes using microoalage in the field and in laboratory culture. Since the partially treated waste water of Orissa Sponge Iron Limited, Palasponga commissioned since 1983 discharge liquid waste containing excessive amount of iron, sulphate, carbonate and suspended particles of coal, iron ore and sponge iron to the surrounding area, the work has been presented here on:

    i.   Occurrence of various micro-algae in the effluent polluted area and its comparison with flora of unpolluted fields of the locality.

    ii.   Comparison of the per cent survival of two heteocystous blue green algae, one isolated from the effluent polluted and the other from unpolluted fields on various concentrations of the waste water.

    iii.   Growth characteristics of the test organisms in iron rich and iron deficient media in the laboratory cultures.

3.3.1 The study site

Orissa Sponge-iron Ltd. (21° 52′ N, 85° 38′ E) is located at Palasponga, 20 km away from Keonjhar district headquarters (Fig. 8). It was commissioned in March 1983 adopting ACCAR process with an installed capacity of 150,000 M.T. per year. Iron-ore, coal, H.S.D. oil, lime stone and dolomite are the chief raw materials used for sponge iron making. In the process, iron-ore and coal are fed continuously into a rotary reactor along with H.S.D. oil. By this oxygen of the iron-ore is extracted. As a result there are voids in the product and the microstructure appears spongy and hence the name of the product is sponge iron. The environmental hazards originate mainly due to (i) production of dusts at the time of iron- ore and coal preparations, (ii) gases emerging after reduction such as CO_2, CO, NO_2, SO_2 and (iii) discharge of waste waters heavily loaded with dust particles of iron ore, sponge iron and oil (Fig. 9a).

The black coloured liquid waste also contained excessive amounts of sulphate, carbonate, iron and sodium salts (Table-10). The partially treated effluent (after storage in the settling tanks for few days), discharged to the surrounding fields through a outlet of its boundary wall (Fig. 9b, c) which produce thick white crust in a large area. In the rainy season, the waste water flows in all directions and enter into the surrounding rice-fields (Fig. 8 and 9).

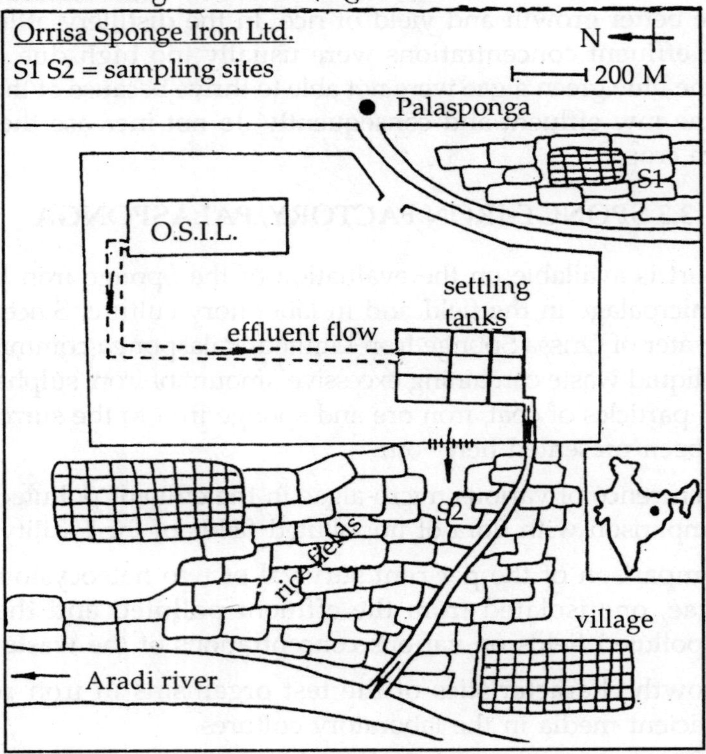

Fig.8: Study site showing location of Sponge-iron factory (Orissa Sponge-Iron Ltd.), Palasponga, keonjhar and the passage of water water from the discharge point of the factory to Aradi river.

Table 10 : Physico-chemical characteristics of the effluent of sponge-iron factory of Palasponga. Waste water was collected from the effluent discharge point. All values except pH and temperature are in mg/litre

Colour	Black
Odour	Burnt ash
PH	6.5
Temperature (° C) at the discharge point	35.0
Total solids	2890 (mg/L)
Dissolved oxygen	0 (mg/L)
Total nitrogen	15.2 (mg/L)
Phosphate	2.6 (mg/L)
Sulphate	174.8 (mg/L)

Chloride	336.4 (mg/L)
Calcium	121.6 (mg/L)
Magnesium	29.2 (mg/L)
Iron (total)	352.0 (mg/L)
Oil and grease	22.0 (mg/L)

3.3.2 Materials and Methods

Soil samples and algal patches were collected from the effluent affected and unaffected fields in every season up to two years (October 1987 to September, 1989). Soil samples were incubated in Allen and Arnon's medium (1955), with or without combined nitrogen, at room temperature under fluorescent light. The algal species appeared in the cultures were identified. Two heteocystous blue green algae, *Calothrix castelli* and *Anabaena torulosa*, the former isolated from the effluent polluted fields and the latter from unpolluted fields of Palasponga, were grown in the laboratory in Allen and Arnon's nitrogen free medium (1955) at 28±2° C under 7.5 W/m² light intensity. Both the organisms were incubated in agar plates in presence of various concentrations of the effluent (v/v, prepared in AA medium) and the survival of the organisms were calculated taking the colonies in control plates as 100 per cent.

Fig.9 a-c: (a) The sponge-iron factory (Palasponga, Keonjhar) and its surrounding area showing the settling basin where the liquid effluents are stored. Arrow indicates the canal through which effluent is carried and discharged outside the boundary of the factory; (b) The effluent canal which emerges from the boundary of the factory and passes through the surroundings fields; (c) Fields in the surrounding area (sampling sites) which receive the untreated effluents from the canal.

Growth experiments were conducted in 100 ml capacity cotton stoppered conical flasks containing 25 ml of medium. For iron deficient culture, Fe-EDTA was omitted from the culture solution. pH of the medium of all experimental cultures were adjusted to 7.5 prior to sterilization. Growth was recorded by light scattering of the algal suspension at 760 nm following the method described earlier.

3.3.3 Results and discussion

Occurrence of different algal forms in the Sponge-iron factory effluent affected and unaffected fields of Palasponga for the period from October 1987 to September 1989 are presented in Table-11.

Table 11 : Occurrence of different algal forms in the sponge iron factory (OSIL, Orissa) effluent affected and unaffected fields of Palasponga.

Organism	Effluent affected field	Effluent unaffected field
Aphanothece saxicola	-	+
Gloeothece sp.	++	+
Oscillatoria anomalis	+	+
Lyngbya major	++	+
Nostoc piscinale	-	+
Anabaenopsis circularis	+	-
Anabaena variabilis	-	+
Anabaena torulosa	-	++
Nodularia spumigena	+	-
Calothrix marchica	+	+
Calothrix castelli	++	-
Scenedesmus dimorphus	+	-
Pinnularia subscapitata	+	-
Surrella linearis	+	-
Pediastrum sp.	+	+

+, occur; ++, occur abundantly

Though 15 species were recorded, mostly ensheathed forms of blue green algae, viz. *Gloeothece, Lyngbya* and *Calothrix* and few green algal forms were found growing in the polluted fields. Experiments were conducted to find out the comparative tolerance of *Calothrix castelli* and *Anabaena torulosa,* commonly occurring in effluent polluted and unpolluted fields respectively in the presence of various concentrations of the Sponge-iron factory effluent. It was observed that *Calothrix castelli,* which has a sheath around its trichome, could tolerate up to 50%

(v/v) of the effluent whereas *Anabaena torulosa*, isolated from unpolluted rice-fields, could not survive in presence of >30% (v/v) of the waste water (Fig. 10). Since the waste water contain excessive amount of iron, experiments were conducted to find out the growth characteristics of both the organisms in the presence or absence of iron in the culture solution. It was observed that though growth of the organisms were not significant in the iron deficient medium in comparison to the control (Fig. 11), presence of excess iron (0.2% Fe-EDTA, w/v) or the effluent (50%, v/v) decreased the growth of both the blue green algae; reduction in growth was more significant in *Anabaena torulosa*, which do not have sheath around its trichome.

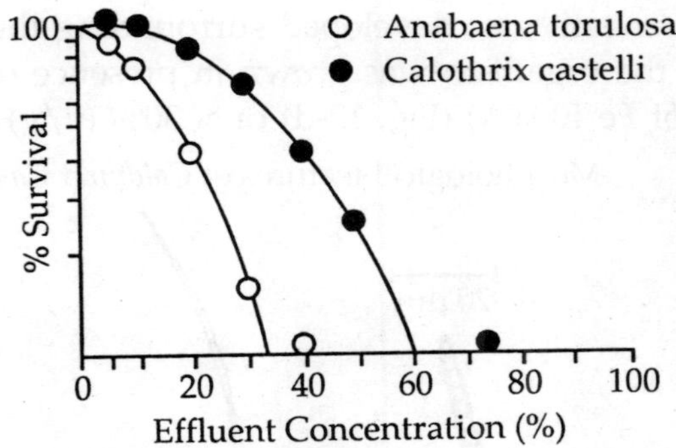

Fig.10: Per cent survival of *Calothrix castelli* and *Anabaena torulosa* isolated from Sponge-iron factor efflluent polluted and unpolluted fields respectively, in presence of various concentrations of the effluent (prepared in AA medium).

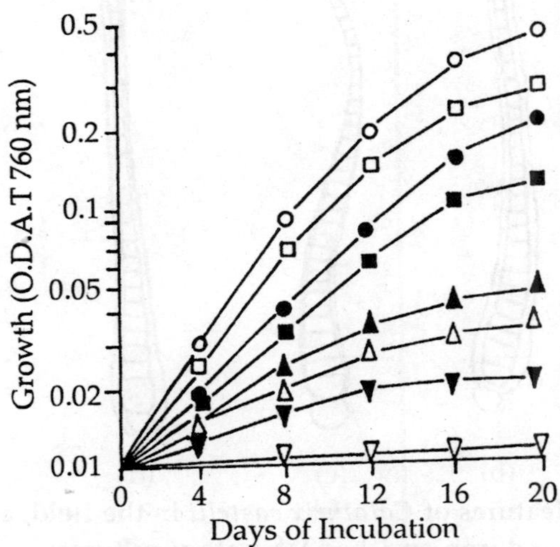

Fig.11: Growth of *Calothrix castelli* and *Anabaena torulosa* in iron deficient (-Fe EDTA), iron enriched (+ 0.2% Fe EDTA) and Sponge-iron factory effluent (50%, v.v) supplemented medium. *Anabaena torulosa*: ○, control; □, iron deficient; △, iron enriched; ▽, effluent supplemented culture. *Calothrix castelli*: ●, control; ■, iron deficient; ▲, iron enriched; ▼, effluent supplemented culture.

Morphological features of *Calothrix castelli* in presence or absence of iron in the culture medium (Fig. 12) throw some light on the survival of the organism on higher concentration of iron and the effluent. Trichomes of *C. castelli*, collected from polluted fields were short and surrounded by a thick sheath layer. The heterocyst was found compressed within the sheath and a spore was observed adjcent to the heterocyst (Fig.12-a). When the organism was isolated and grown in Allen and Arnon's nitrogen free medium, elongated trichomes, enclosed by fairly thick sheath was seen (Fig. 12-b). In the iron deficient (-Fe EDTA) culture, the trichomes were much elongated and tappered into vacuolated and hairy cells. In this case a thin sheath was found surrounding only the basal portion of the trichome(Fig.12-c). However, a thick sheath was developed surrounding the trichome, including the heterocyst, when the organism was grown in presence of higher concentration of iron (0.2%, w/v of Fe EDTA) (Fig. 12-d) or > 30% (v/v) of the effluent (Fig.12-e).

Morphological features of Calothrix Castelli

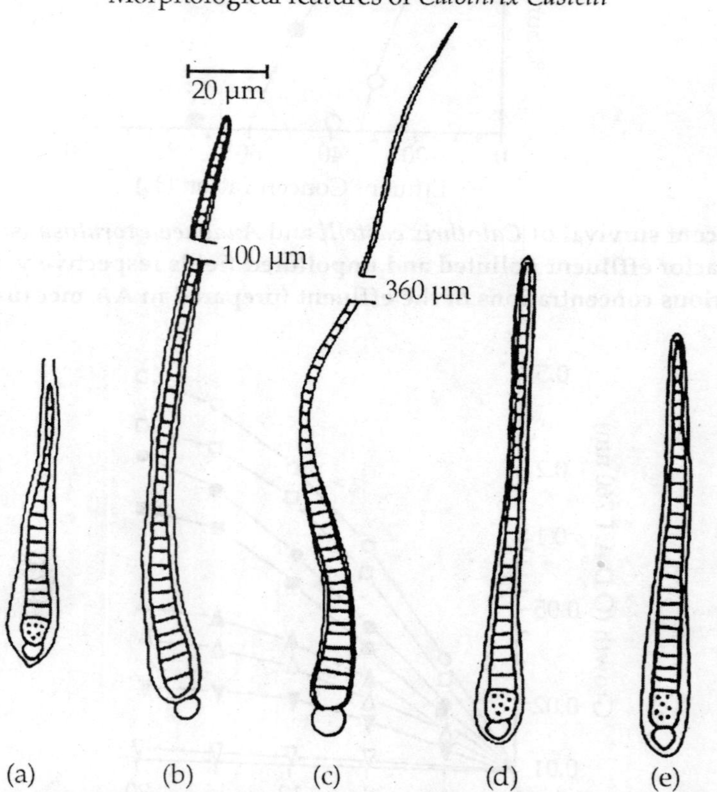

Fig.12: Morphological features of *Calothrix castelli* in the field, and in iron deficient and iron enriched laboratory cultures:

(a) in the effluent polluted fields, (b) in Allen and Arnon's nitrogen free medium, (c) in the iron deficient (-Fe EDTA) AA medium, (d) in the iron enriched (+ Fe EDTA, 0.2%, w/v) AA medium, (e) in presence of 50% (v/v) effluent in AA medium.

These results show that the waste water of the Orissa Sponge-iron limited is harmful for growth of a variety of blue green algae In rainy season, the effluent enter into the surrounding rice-fields and thus there is a possibility of their effect on the retardation of growth of a variety of nitrogen-fixers in the rice field environments. Most of the blue green algal forms occur in the polluted fields has well defined sheath. Only the ensheathed forms of blue green algae were found tolerating high concentrations of industrial effluents in laboratory culture. Thus it is fairly convincing that these outermost surface structures play an important role for making ensheathed forms of blue green algae to thrive in the adverse conditions.

3.4 FERTILIZER FACTORY (FCI), TALCHER

The fertilizer factory of Talcher (Fertilizer Corporation of India, Talcher unit) discharges 1-3 million gallons of partially treated or untreated effluents per day into Nandira River. A number of villages and rice-fields are located on the river bank. During the rainy season, waste water enters into the surrounding rice-fields and destroys the crop. Thus it is an urgent concern to analyse the physico-chemical characteristics of the waste water discharged from the factory and to evaluate the extent of pollution caused by the waste water using a nitrogen-fixing blue green alga of local unpolluted rice fields as the test organism.

3.4.1 The study site

The fertilizer factory, Talcher (20° 57′ N, 85° 15′ E) has manufactured urea and ammonia since 1979. The factory discharges effluent to Nandira River through a 3-km long open canal (Fig.13, 14a,b). Physico-chemical analysis of the effluent was conducted following the methods given in APHA (1989). The effluent was black in colour, highly alkaline and contained objectionable amounts of total solids, ammoniacal nitrogen and oil. Survey of the polluted area for the occurrence of various algal species and zooplankton, mostly at three sampling sites (S 1, point of discharge of the effluent from the factory; S 2, the site where effluent canal mixes with Nandira River and S 3, 2 km away in the down stream from the above confluence point) (Fig. 13) were carried out throughout the years 1988 and 1989.

3.4.2 Materials and methods

The nitrogen-fixing blue green alga, *Westiellopsis prolifica* Janet occurring predominantly in the rice-field soils of Orissa (Adhikary and Pattnaik 1979) was employed as the test organism to evaluate the nutrient status of the fertilizer factory effluent. Experiments were conducted using axenic cultures of *W. prolifica* grown in Allen and Arnon's nitrogen free medium (1955) with the micronutrients as used by Fogg (1949) at 28±2° C and 7.5 W/m² light intensity.

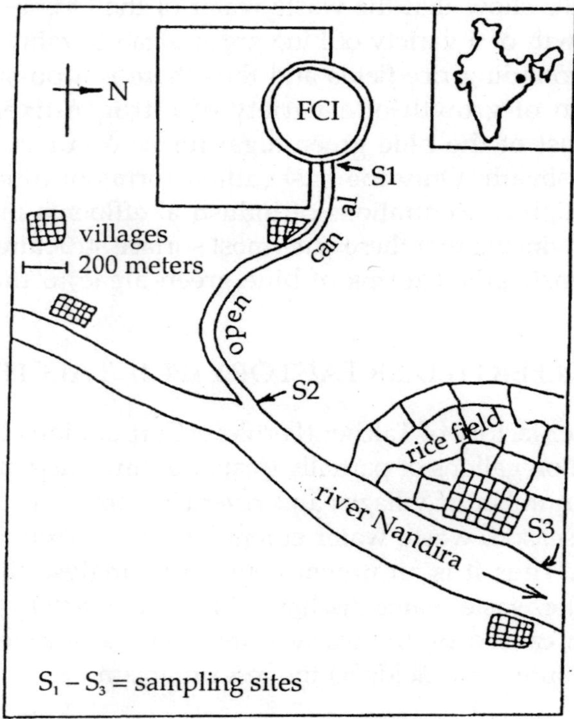

Fig.13: Study site showing location of the Talcher Fertilizer factory (Fertilizer Corporation of India) and the passage of waste water from the discharge point of the factory to Nandira river.

Growth was estimated by measuring absorbance of the culture suspension at 760 nm by Spekol (Carlzeiss) spectrophotometer. Chlorophyll and carotenoid pigments were extracted with 80% acetone and quantitatively determined following Mackinney (1941) and Davies (1976) respectively. Change in morphological features and survivability of the organism was studied in agar plates containing varying concentrations of the industrial effluent (1, 5, 10, 20, 30, 40, 50, 70 and 100%, v/v), prepared in Allen and Arnon's nitrogen free medium. Aliquots of 1.0 mL diluted suspension of the organism was spread aseptically on each of the agar plates and incubated under light in the culture room. After 10 d of growth, colonies were counted and survival data were calculated taking survival on control plates as 100%. Total nitrogen content of certain cultures were estimated following Herbert *et al.* (1971). The amount of nitrogen fixed was determined after substracting the nitrogen present in the medium and inoculum.

3.4.3 Results and discussion

The fertilizer factory at Talcher discharges millons of gallons of untreated waste water every day into Nandira River. The effluent was highly alkaline (pH >

10) and contained excessive amounts of total solids, oil and ammoniacal nitrogen (Table-12). No algal growth was visible either at the effluent discharge point (site 1) or in the open canal which mixes with Nandira River (site 2). In the down stream (site 3) (Fig. 14-d) only a few blue green and green algae, e.g. *Gloeocapsa, Chroococcus, Lyngbya, Oscillatoria, Chlorella, Scenedesmus, Coelosphaerium, Cylindrocystis, Hydrodictyon* and *Melosira* appeared in the river bed (Table-13). Most of these organisms are found to possess a well defined sheath around their cells/trichome. Such ensheathed forms of blue green algae which have also been reported from various other industrial waste polluted water bodies (Adhikary 1987; Adhikary and Sahu 1988), might be tolerant of the waste water due to possession of such adaptive morphological features. However, none of the heterocystous blue green algae were recorded either in the river Nandira or in the adjoining rice fields which are periodically over-flooded with the effluent mixed water. Thus to evaluate the toxic effect of effluents on the nitrogen economy of rice-fields, the nitrogen-fixing blue green alga, *Wastiellopsis prolifica*, commonly occurring in this region (Adhikary and Pattnaik, 1979) was used as the test organism.

Fig. 14: (a) The Fertilizer factory (F.C.I., Talcher, Orissa) and the discharge point of untreated waste water of the factory (sampling site no. 1). (b) The waste water is carried out from the factory to Nandira river through the open canal. (c) The mixing point of the effluent with the water of Nandira river (sampling site no.2). Due to high amount of suspended black solid particicles and oily content of the effluent, up to a km distance in the down stream there is a clear separation of the flow of river water and the effluent. (d) Growth of various algal forms forming thick scum in the water in the river bank about 3 km away from the effluent and river water mixing point in the down stream (sampling site no. 3).

Table 12 : Physico-chemical characteristics of Talcher fertilizer factory (FCI, Talcher, India) effluent. The waste water was collected from the effluent discharge point of the factory and from Nandira River at site 1 and site 2, respectively. All values except pH and temperature are in mg/L.

Characteristics	Effluent from discharge point of the factory (site 1)	Effluent mixed with Nandira river water (site 2)
Colour	Blackish	Blackish
Odour	Ammoniacal	Ammoniacal
pH	10.5	10.0
Temperature (at site 1), ° C	48.0	35.0
Total suspended solids (mg/L)	2260.0	1830.0
Biochemical oxygen demand (mg/L)	91.5	82.0
Chemical oxygen demand (mg/L)	520.0	220.0
Dissolved oxygen (mg/L)	2.2	2.8
Ammoniacal nitrogen (mg/L)	850.0	590.0
Nitrate nitrogen (mg/L)	105.0	26.0
Urea (mg/L)	590.0	480.0
Phenol (mg/L)	156.0	90.0
Oil and grease (mg/L)	30.0	22.0
Chloride (mg/L)	360.0	280.0
Sulphate (mg/L)	168.0	148.0
Calcium (mg/L)	35.0	44.5
Magnesium (mg/L)	15.0	24.0
Phosphorus (mg/L)	4.5	4.5

The blue green alga could not survive in presence of more then 50% (v/v) neutralized waste water of the industry. Normal filaments of the organism were found in presence of up to 10% (v/v) of the waste water in the culture medium. With the increase in concentration of the effluent in the medium, heterocysts occur rarely, the cells of the prostrate filaments became enlarged and the number and length of erect branches decreased considerably. With further increase of the effluent proportion in the culture, filamentous structure of the organism was lost, cells became clumped, disorganized and bleached (Table-14).

Table 13 : Occurrence of different organisms in the area polluted by the effluent of Talcher fertilizer factory (FCI, Talcher, India) for the period from November 1987 to March 1989.

Organism	Sampling sites		
	S 1	S 2	S 3
Algae:			
Chroococcus minutes	-	+	++
Gloeocapsa sp.	-	+	++
Lyngbya majuscula	-	+	++
Oscillatoria formosa	-	-	+
Coelosphaerium sp.	-	-	+
Cylindrocystis sp.	-	-	+
Scenedesmus quadricauda	-	-	+
Chlorella pyrenoidosa	-	-	+
Hydrodictyon reticulatum	-	-	+
Melosira granulata	-	-	+
Protozoa: *Glaucoma* sp.	-	+	++
Rotifera: *Platyias* sp.	-	+	+
Copepoda *Cyclops* sp.	+	+	-

In liquid culture, *W. prolifica* grew and fixed nitrogen appreciably in the basal nitrogen free medium up to 20 d (Fig. 15 a,b). The growth rate was further enhanced when the medium contained 10% (v/v) of the fertilizer factory effluent. But when the basal medium was supplemented with KNO_3 or 50% (v/v) of the effluent, growth was inhibited. However, nitrogen-fixing capacity of the organism was decreased over control even in presence of 10% (v/v) of the effluent (Fig. 15 b). It is well known that growth of nitrogen fixing blue green algae are suppressed or reduced by supplementing combined nitrogen into the culture medium (Fogg *et al.* 1973). Reduction in growth and nitrogen fixing capacity and supperession of heterocysts and erect branches in presence of higher concentrations of the waste water proves that the effluent discharged outside of the factory is nitrogen rich.

Table 14 : Effect of different concentrations of the effluent* of the Fertilizer Factory (FCI, Talcher) on the survival and morphological features of *Westiellopsis prolifica*. Culture were grown at 28±2° C under 7.5 W/m² light intensity up to 15 days

Effluent concentration (%, v/v)	Survival ** (%)	Heterocyst	Branching	
			Prostrate filaments	Erect branches
0 (control)	100	4.2±0.8***	Barrel shaped cells, as long as broad ****	Profuse branching

Effluent concentration (%, v/v)	Survival ** (%)	Heterocyst	Branching	
			Prostrate filaments	**Erect branches**
1	100	Few in number	Barrel shaped cells, as long as broad	Profuse branching
5	100	Few in number, granular	Barrel shaped cells, as long as broad	Reduced in length
10	98	Few in number, granular	Barrel shaped cells, as long as broad	Reduced in number and length
20	85	Rarely occur	Cells enlarged and granular	Few in number and suppressed
30	62	Rarely occur	Cells enlarged and granular	Few in number and suppressed
40	45	Suppressed	Cells enlarged, with branch initials	Suppressed
50	12	Suppressed	Cells enlarged, with few branch initials	Absent
70 and 100	0	No filamentous structure	Cells clumped and bleached	

*   Different concentrations of the effluent were prepared in basal inorganic medium and adjusted to pH 7.5 by aseptic addition of N/10 HCl.

**   Survival in presence of different concentrations of the effluent was computed taking the survival in the control cultures as 100.

*** Frequency of heterocysts (number of cells between two successive heterocysts).

**** Barrel shaped cells which are about as long as broad (14.6 - 16.1 µm long and 11.0 - 12.1 µm broad)

It was also observed that chlorophyll content of *W. prolifica* increased in presence of 10% (v/v) effluent in the culture medium over control (Fig. 16 a). But in presence of KNO_3 or higher concentrations of the effluent (50%, v/v) in the medium, chlorophyll content per cell decreased significantly and after 12 d of incubation, cells became colourless. Decrease in the carotenoid content of the cells in presence of combined nitrogen sources followed a similar trend but the degradation was slow in comparison with the chlorophyll value (Fig.16 b). This shows that the nitrogen rich fertilizer factory effluent affects the photosynthetic system by degrading the pigment content of the cells thereby making the photosynthetic organisms unable to thrive in the waste water of the industry.

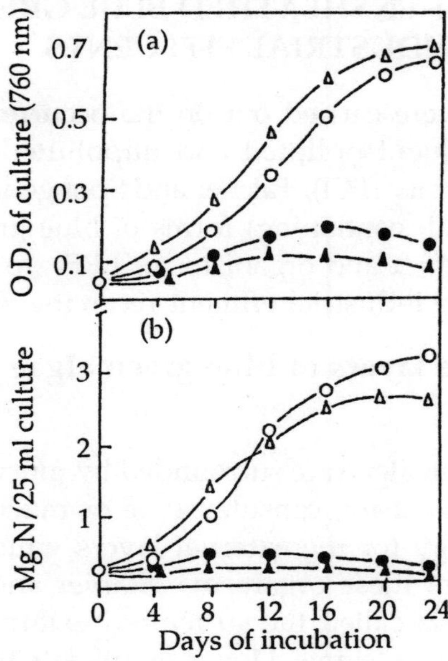

Fig.15: Growth (a) and nitrogen-fixation (b) of *Westiellopsis prolifica* in presence of Fertilizer factory effluent. ○,control (c); ●, c+ 2g/l KNO_3; Δ, c+10% (v/v) effluent; Δ,c+ 50% (v/v) effluent.

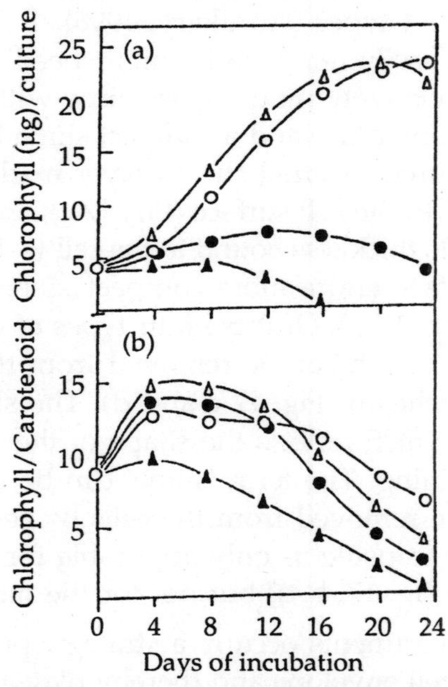

Fig. 16: Chlorophyll a content (a) and chlorophyll- carotenoids ratio (b) of *Westiellopsis prolifica* in presence of fertilizer factory effluent. ○, control (c); ●, c+ 2g/l KNO_3 ; Δ c+ 10% (v/v) effluent; Δ, c+ 50% (v/v) effluent.

3.5 TOLERANCE OF ENSHEATHED BLUE GREEN ALGAE TO INDUSTRIAL EFFLUENTS

Ecological studies were carried out on the occurrence of blue green algal forms in the industrial effluent polluted and unpolluted areas of the Molasses distillery, Aska, Fertilizer factory (FCI), Talcher and Sponge-iron factory, Palasponga. Only few ensheathed (sheath containing) forms of blue green algae belonging to *Chroococcus, Gloeothece, Calothrix* and organisms of LPP group (*Lyngbya-Plectonema-Phormidium*) occurred in the industrial effluent receiving areas.

Which external envelope layers of blue green algae has been designated as sheath?

The cells of blue green algae are surrounded by mucilaginous external layers which have been named as sheath, capsule, slime or mucilage. Until 1980s, there was no uniform terminology for the external layers which surround the gram-negative type of cell wall of these organisms (Stanier and Cohen-Bazine, 1977). Drews and Weckesser (1982) called the structured external layer a 'sheath' and the unstructured layer/zones as 'slime'. However, till date these terminologies have not been strictly followed. Therefore, structural and functional differences among these layers are not always evident in the literature. Based on the literature, the different types of external layers surrounding the cell wall of blue green algae have been recently grouped into 4 types (Adhikary, 1998). 1. Thin slime layer around the cell wall which can be easily removed by high speed centrifugation of the cell suspension (Fig. 17 a), 2. relatively more dense slime with two apparent layers: a diffused slime layer in the outer face and a compact slime layer adhering to the cell wall (Fig. 17 b), 3. structured external sheath layer made up of fibrous material with an orientation parellel to the cell surface (Fig. 17 c), and 4. structured external sheath layer of considerable thickness bound to the cell wall with a loosely arranged fibrous layer at the inner face and a more compact, densely packed fibres in the outer area of the sheath (Fig. 17 d). Of these four types of external envelope layers, the rigid structured layers which can be removed from the cells only after harsh treatment can be called as sheath (Fig. 17 c and d). The sheath is a uniform layer surrounding the cell wall which reflects the shape of the organisms and is visible even in the absence of staining. The term 'slime' can be used for the extracelluar diffuse layers which can be removed from the cells by centrifugation [Fig. 17 (a.2 and b.3)]. The terminology 'capsule' is only applicable for the compact slime layer adhering to the cell wall [(Fig. 17; b.4)] but not for the other layers.

The extracelluar investments occupy a strategic position at the outer most region of blue green algal cell envelope and thereby play a fundamental role in the interaction of cells with their environment (Costerton *et al.*, 1987). With a view to understand the higher tolerance capacity of ensheathed blue green algae than sheathless forms in the field, growth response of different types of ensheathed blue green algae, viz. *Gloeothece* PCC 6909 (Fig. 17 d type), *Calothrix marchica* var. *intermedia* and *Westiellopsis Prolifica* and sheathless *Anacystis nidulaus* (CCAP 1405/1) to various concentrations of neutralized distillery, fertilizer factory and sponge-iron factory effluents was studied in the laboratory. *Calothrix parietina* and *Westiellopsis prolifica* were selected for this investigation because they grew well in the rice fields of this region in the east-coast of India and possess higher nitrogen fixing capacity (Tripathy *et al.*, 1990). Experiments were also conducted to findout the chemical nature of the isolated sheath fractions of *Gloeothece* PCC 6909 grown in presence or absence of the industrial effluents.

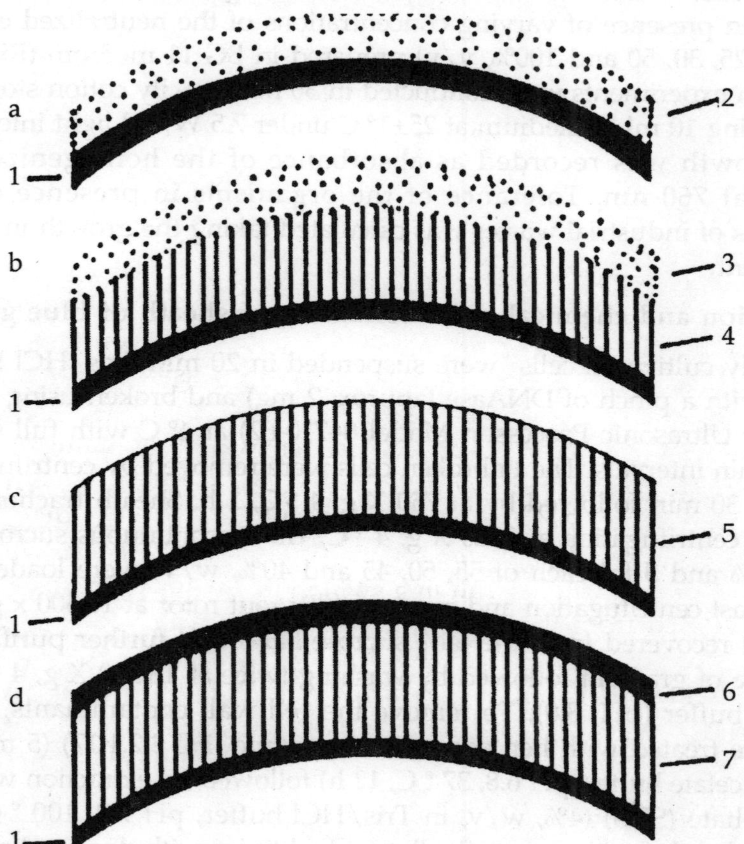

Fig. 17: Schematic diagram of different types of external layers (a-d) surrounding gram -ve cell wall of cyanobacteria (1) cell wall; (2) slime layer loosely bound to cell wall; (3) diffuse slime layer in the outer face and (4) compact slime layer adhering to cell wall; (5) structured external layer bound to cell wall;
(6) structured external layer with a compact layer in the outer face and (7) loose layer adhering to cell wall in the inner face. Examples: (a) *Synechococcus* PCC 6803, PCC 6714; *Aphanocapsa halophytia* MN-11 type; (b) *Microcystis flos-aquae* C 3-40 type; (c) *Chroococcus minutus* SAG B. 41.79 type; (d) *Glocothece* PCC 6909 type.

3.5.1 Materials and methods

3.5.1.1 Growth experiments

Four different species of blue green algae, (i) *Calothrix marchica* Lemm. var *intermedia* Rao, isolated from the distillery effluent polluted area, (ii) a nitrogen fixer in the rice fields of this region, *Westiellopsis prolifica* Janet, (iii) the unicellular *Anacystis nidulaus* (CCAP 1405/1) obtained from Cambridge Culture Collection of Algae and Protozoa, and the colonial ensheathed *Gloeothece* PCC 6909 obtained from Pasteur Culture Collection were used as the experimental organisms. All these organisms were grown in presence of varying concentrations of the neutralized effluents (1, 5, 10, 15, 20, 25, 30, 50 and 100%, v/v) prepared in BG 11 medium (Rippka *et al.*, 1979). Growth experiments were conducted in 30 ml capacity cotton stoppered test tubes containing 10 ml of medium at 25±1° C under 7.5 W/m² light intensity up to 20 days. Growth was recorded as absorbance of the homogenized culture suspension at 760 nm. Tolerance of the organisms in presence of various concentrations of industrial wastes was calculated taking the growth in the control as 100 per cent.

3.5.1.2 Isolation and chemical characterization of sheath of blue green algae

Freshly cultivated cells were suspended in 20 mM Tris/HCl buffer (pH, 8.0), mixed with a pinch of DNAase (approx. 2 mg) and broken using a sonicator (Cole-parmar Ultrasonic Processor, Model 04710-62) at 4° C with full speed for 4 hours at 10 min intervals. The unbroken cells were removed by centrifuging at 121 X g, 4° C for 30 min followed by at 750 X g, 4 ° C, 1 h. Sheath fraction was then separated by centrifugation at 3020 X g, 4 ° C, 1h. Discontinuous sucrose gradient (10 ml of 60% and 5 ml each of 55, 50, 45 and 40%, w/v) were loaded with the sediment of last centrifugation and run in a swingout rotor at 16,300 x g, 4° C, 4 h. Sheaths were recovered from the 60% sucrose band and further purified once in the same type of gradient followed by washing twice at 16,300 X g, 4 ° C, 30 min in Tris/HCl buffer (pH, 8.0). To remove the cell wall contaminants, the sheath fractions were treated with hen egg white lysozyme (EC 3.2.1.17) (5 mg in 25 ml ammonium acetate buffer, pH 6.8, 37 ° C, 12 h) followed by extraction with sodium dodecyl sulphate (SDS) (4%, w/v, in Tris/HCl buffer, pH 7.8, 100 ° C, 15 min). The purified sheath fraction were finally washed twice with double distilled water (16,300 X g, 4 ° C, 30 min), dialysed, collected by centrifugation (16, 300 X g, 4° C, 1 h) and freeze dried.

Freshly harvested cells or isolated sheath fractions were fixed in 1% (w/v) osmium tetroxide, dehydrated in alcohol and polymerized in Epon according to the procedure followed by Golecki (1977). Ultra-thin sections were mounted on Formvar-coated copper grids and stained with uranyl acetate (Watson, 1958) followed by lead citrate (Reynolds, 1963). Specimens were examined in a Philips EM 400 microscope at 80 kV. Light microscopic photograph of the cells and sheath fractions were taken in a Meiji ML-TH-05 Trinocular phase contrast microscope using Agfapom 100 film.

3.5.1.3 Chemical analysis

Neutral sugars were liberated from the freeze dried sheath fraction by hydrolysis with 1 N H_2SO_4 at 100° C for 4 h. The hydrolysates were neutralized with a saturated aquatic solution of $Ba(OH)_2$. Qualitative assay of the neutral sugars in the hydrolysates was carried out by thin-layer chromatography. For TLC on cellulose plates (0.1 mm, Merck, Darmstadt), the solvent n-butanol-pyridine-water (6:4:3. v/v/v) was used and were stained with anilinium hydrogen phthalate (Partridge, 1949).

For gas-liquid chromatography, the sheath fraction was hydrolysed in 0.1 N HCl at 100° C for 48 h. The aldoses of the neutralized hydrolysates (with Amberlite IRA 410/H Co_3^-) were converted to alditol acetates and used for analysis in a Aimil-Nucon microprocessor based (5765 model) gas chromatogaph with FID fitted with a glass column (0.32 by 152 cm; filled with ECNSS-M, 3% on Gas-Chrom Q, 100 to 120 mesh) at a column temperature 180° C, injector and detecter temperature 230° C, carrier gas N_2 flow 30 ml/min. Inosit was used as the internal standard.

Fatty acids liberated by hydrolysis of the sheath fraction in 4 N HCl at 100° C for 6 h, were identified as their methyl ester derivatives by gas-liquid chromatography. Tridecanoic acid (C-17) was added as an internal standard. For identification, Aimil-Nucon 5765 gas liquid chromatograph with FID fitted with a stainless-steel column (0.32 by 150 cm; filled with EGSS-X, 15% on Gas-chrom P, 100 to 120 mesh) at a column temperature 180° C (isotherm), injector and detecter temperature 230° C, carrier gas N_2 flow 30 ml/min was used.

Aminoacids were liberated by hydrolysis of the sheath fraction in 6N HCl at 100° C for 16 h. HCl was removed from the hydrolysate by repeated evaporation in vacuo in presence of KOH in a desiccator. Quantitative determination of amino acids was performed using ninhydrin at 670 nm following the modified procedure of Rosen (1957). For certain experiments (analysis of sheath of *Gloeothece* PCC 6909 in BG 11 medium is the absence of any industrial effluent), amino acids and amino sugars of the hydrolysate was quantitatively determined in an automatic aminoacid analyser (Durrum, model D-500). Quantitative determination of organic phosphorus was carried out following Lowry *et al.* (1954). Uronic acids were quantitatively determined by colorimetric carbazole assay at 530 nm (Galambos, 1967).

3.5.2 Results and discussion

3.5.2.1 Tolerance of various blue green algae to industrial effluents

Four blue green algae of which three ensheathed forms viz. *Gloeothece* sp. PCC 6909, *Calothrix marchica* and *Westiellopsis prolifica* and one sheath less type,

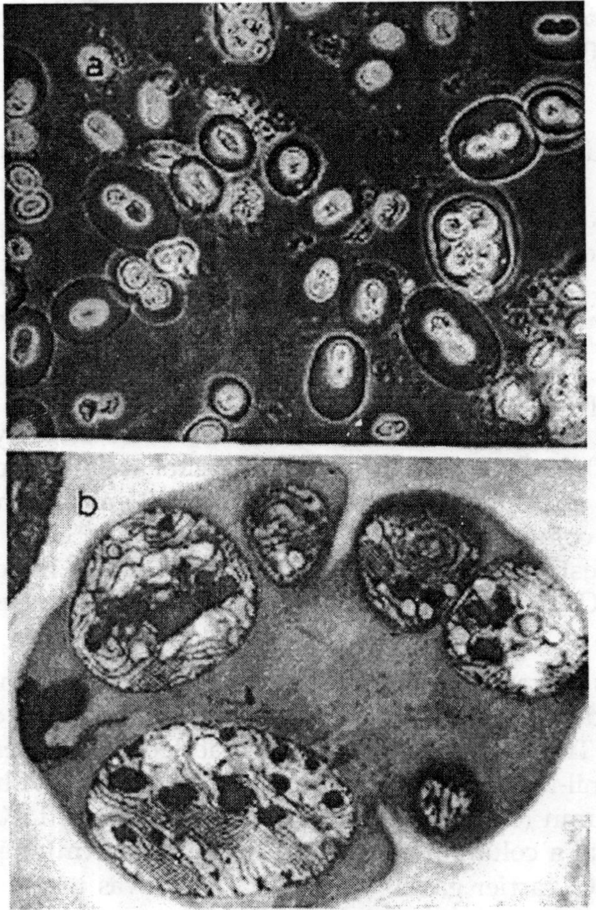

Fig. 18: (a) Light microscopic photograph of *Gloeothece* PCC 6909 cells showing a thick sheath layer (x 1,050). **(b)** Ultrastructure of *Gloeothece* PCC 6909 showing the cells surrounded by well defined sheath around the cell wall (x 9,000).

Anacystis nidulans CCAP 1405/1 were grown in presence of various concentrations of neutralized distillery, sponge-iron factory and fertilizer-factory effluents to understand the cause of occurrence of predominantly ensheathed forms of blue green algae in industrial waste water polluted areas. Of these organisms, *Gloeothece* PCC 6909 showed single cells and cell aggregates surrounded by distinctly visible sheath layers (Fig. 18a). In electron micrograph, several of the cells not connected with each other found embedded in a well developed fibrillar sheath; one of the cell showing division by binary fission (Fig. 18 b). *Gloeothece* PCC 6909 possessing a well defined sheath outside the cell wall tolerated and grew well in presence of higher concentrations of the waste water of all the three industries (Fig. 19-21). The ensheathed *Calothrix marchica* tolerated and grew in presence of up to 15% (v/v),

10% (v/v) and 5% (v/v) of molasses distillery, sponge-iron factory and fertilizer factory effluents respectively. To the contrary, the sheathless *Anacystis nidulans* and thin-sheathed *Westiellopsis prolifica* grew in the media containing only up to 10% (v/v) of distillery and 5% (v/v) of sponge-iron factory and fertilizer factory effluents at a comparable rate to that of distillery effluent in the medium (Fig. 19). Further increase in the effluent concentration in the culture medium affected adversely on the growth of both the organisms. The latter two species though could not survive in presence of 30% (v/v) of the effluents, the ensheathed *Calothrix marchica* as well as *Gloeothece* PCC 6909 tolerated up to 50% (v/v) of the waste waters; higher tolerance values was observed by *Gloecothece* PCC 6909 than *C. marchica*. (Fig. 19-21). Similar results on the tolerance of ensheathed blue green alga *Calothrix castelli* to higher concentrations of sponge-iron factory effluent than the thin-sheathed *Anabaena torulosa* was observed earlier (Chapter 3.3.3). These results on the tolerance of ensheathed blue green algae to higher concentrations of industrial waste waters indicate the protective nature of extracelluar sheath of the organisms to such adverse conditions which are not conducive for the growth of sheathless forms.

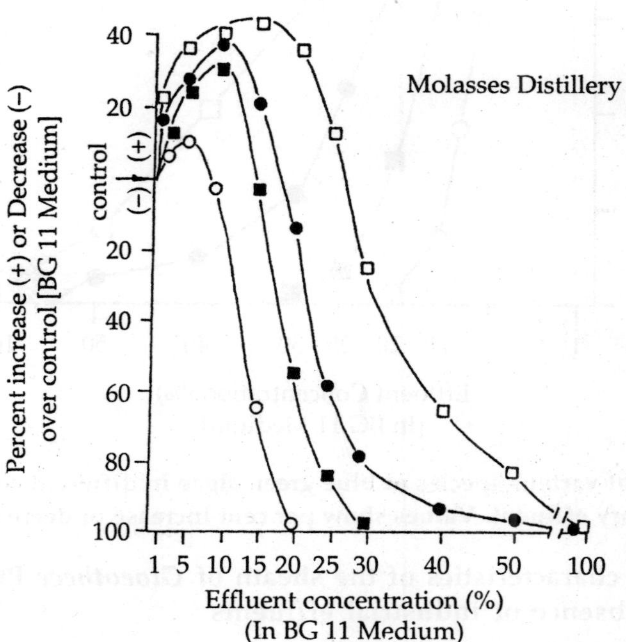

Fig. 19: **Growth of various species of blue green algae in different concentrations of distillery effluent. Values show per cent increase or decrease over control.**

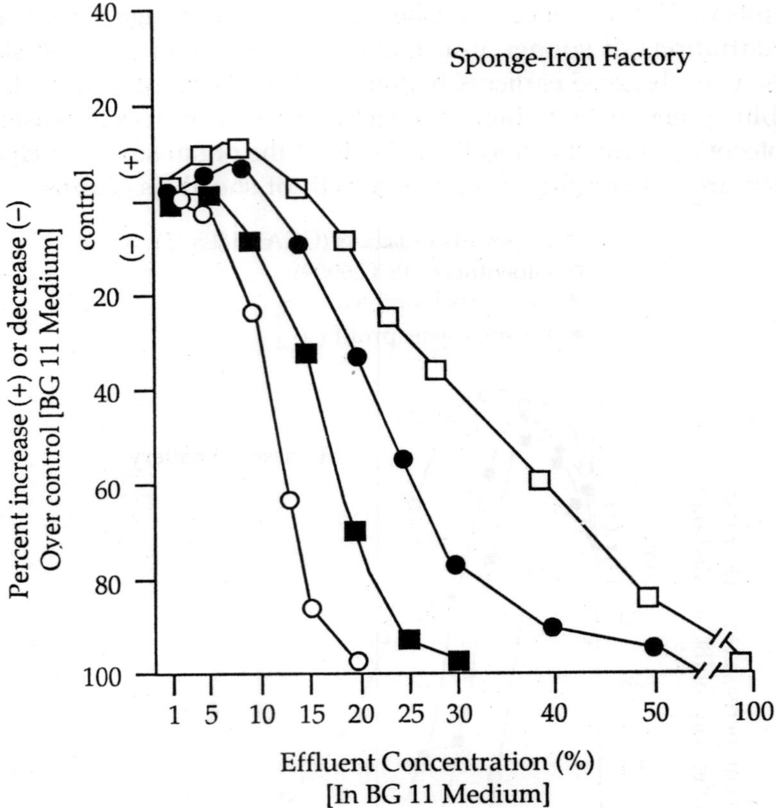

Fig. 20: Growth of various species of blue green algae in different concentrations of Sponge-iron factory effluent. Values show per cent increase or decrease over control.

3.5.2.2 Chemical characteristics of the sheath of *Gloeothece* PCC 6909 grown in presence or absence of industrial effluents

Well defined sheath enclosing both individual cells and cell group of *Gloeothece* PCC 6909 in all the growth conditions was observed by microscopy (Fig. 18). Thus the organism was used for isolation and chemical characterization of the sheath material. The sheath was shed from the cells by mechanical stress, separated from the cell hamogenate by low speed centrifugation and purified by sucrose density gradient centrifugation. Yield of sheath material (Fig. 22 a) of the organism grown in BG 11 medium and from the distillery, fertilizer factory waste and sponge-iron factory water (10%, v/v) supplemented cultures of *Gloeothece* were 8.2, 10.6, 8.0 and 9.1 per cent respectively on a cell dry weight basis. Electron-microscopic photograph of the purified sheath fraction showed a homogeneous fibrillar fine structure (Fig. 22 b), however, the isolated sheath fraction did not represent the complete sheath. The loosening of sheath structure in the isolated sheath compared with the original state was due to severe mechanical treatment applied during isolation. The sheath fraction obtained from the sucrose gradient was composed predominantly of 21.4% (w/w) carbohydrates. Rhamnose, 2-0-methyl pentose, xylose, mannose, galactose, glucose and an unknown sugar was present; galactose, glucose and mannose being the dominant sugars in the sheath material (Table-15). These sugars were detected by TLC and gas liquid chromatographic analysis (Fig. 23 and 24). Lipids, phosphorus, glucosamine, uranic acids were present in trace amounts (1.0, 2.1, 2.4 and 1.9% respectively of the dry weight). A considerable proportion of the sheath consisted of protein. Amino acid residues of the sheath fraction from sucrose gradient accounted for 12.9% of the dry weight (Table-15). After treatment of the sheath fraction with lysozyme and SDS for removal of cell wall contaminants, the glucosamine and uronic acid contents were removed. The protein was not totally solubilized with this treatment, however, the amount of total aminoacids were considerably reduced (8.4%) with an increase in the percentage of sugar content (31.4%) of the sheath material (Table-15, Fig. 25). The purified sheath also contained negligible amount of total fatty acids (Fig. 26) and phosphate. Absence of muramic acid, diaminopimelic acid and β-hydroxy fatty acids, law phosphorus and lipid content suggests that sheath fraction was free from membrane contamination. Quantitative differences was observed in the composition of sheath materials of the organism grown in presence of industrial waste waters (Table-16). Sheath was more voluminous in distillery effluent supplemented cultures and contained 34.7% more sugars and 26% less protein than the sheath of the cells grown in BG 11 medium.To the contrary, though the yield of sheath material from the cells grown in presence of fertilizer factory effluent was almost same as the cells from the basal medium, the former contained 15.6% and 61.9% more carbohydrate and protein respectively than the latter material. Further, sugar and protein content in the sheath of the organism grown in presence of sponge-iron factory effluent was also increased over control by 1.9% and 23.8% respectively. Though the quantity of lipids in the sheath of industrial waste grown cells remained unchanged, phosphorus content was increased slightly and uronic acids was detected in their sheath fractions.

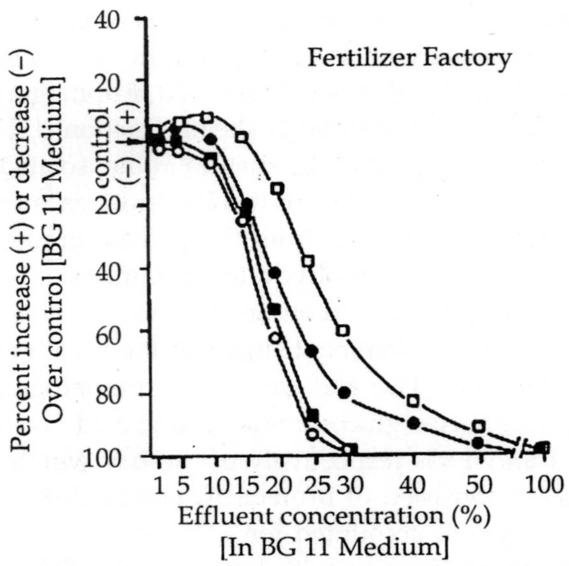

Fig. 21: **Growth of various species of blue green algae in different concentrations of Fertilizer factory effluent. Values show per cent increase or decrease over control.**

Chemical analysis of the slime layers of certain blue green algae have revealed an array of neutral sugars and uronic acids similar to those found in bacterial external layers (Hough *et al.*, 1952; Bishop *et al.*, 1954; Moore and Tischer, 1965; Dunn and Wolk, 1970; Sangar and Dugan, 1972; Wang and Tischer, 1973; Mehta and Vaidya, 1982; Painter, 1983; Plude *et al.* 1991; Gloaguen *et al.*, 1995). The more structurally resilient sheaths of *Chlorogloeopsis* sp. PCC 6912 (Schrader *et al.*, 1982); *Chroococcus minutus* SAG B 41.79 (Adhikary *et al.*, 1986); *Gloeothece* PCC 6501 (Weckesser *et al.*, 1987), *Calothrix parietina* D550 (Weckesser *et al.*, 1988) and *Fischerella* PCC 7414 (Pritzer *et al.*, 1989) have been shown to contain O-methyl sugars and a protein component in addition to the typical sugar residues detected in previously studied external layers. Like the composition of sheath of these blue green algae, the complex sugar spectrum with pentoses, hexoses and O-methyl sugars indicate high specificity of the sheath of *Gloecothece* PCC 6909. The presence of proteins, uronic acids and varying amounts of phosphate and fatty acids in addition to neutral sugars indicate a significant degree of complexity. The presence of a variety of components in the sheath of *Gloeothece* PCC 6909 which quantitatively varies when grown in various industrial effluents may convey their resistance to toxic compounds and to bacterial decomposition, since several enzymes may be needed to achieve complete breakdown of the heteropolysaccharides.

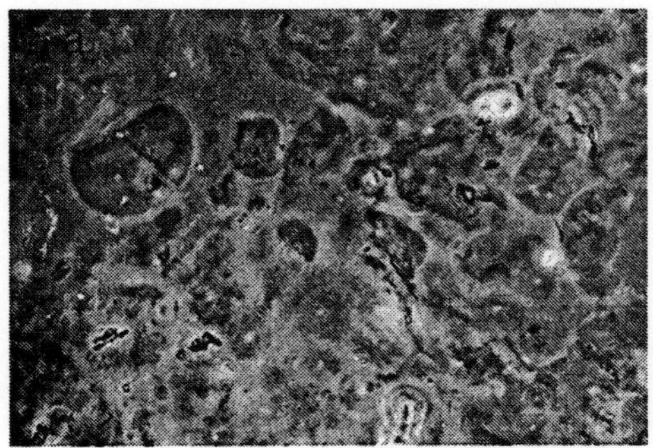 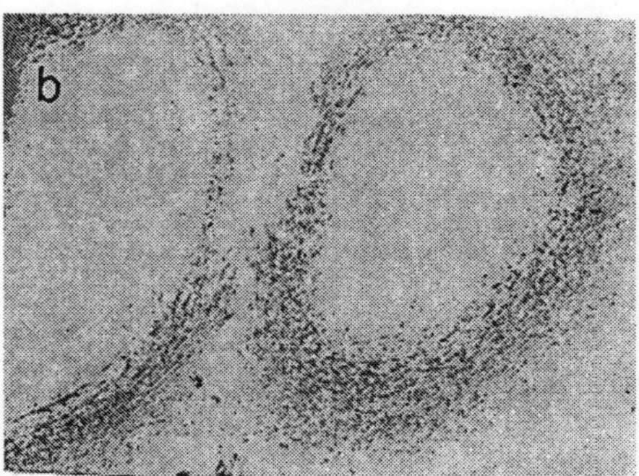

Fig. 22: (a) Light microscopic photograph of isolated sheath material shed from *Gloeothece* PCC 6909 by mechanical means and isolated after sucrose density gradient centrifugation (x 920). (b) Electron microscopic photograph of isolated sheath material of *Gloeothece* PCC 6909 after treatment with lysozyme and hot SDS (x 7,500).

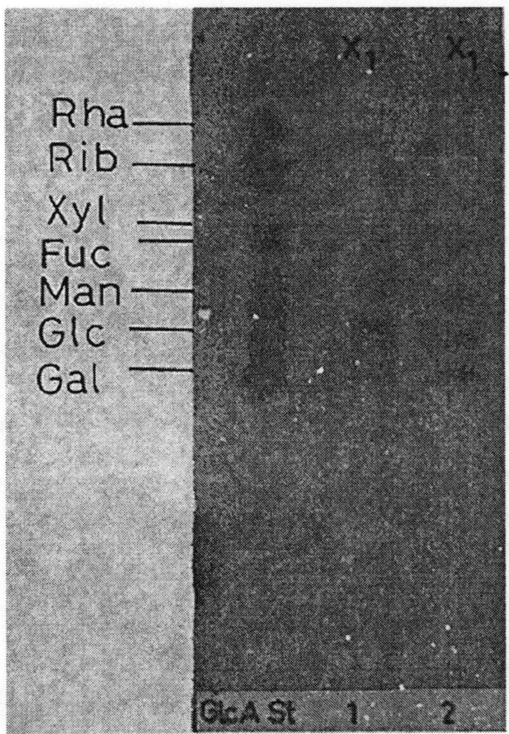

Fig. 23 : Analysis of neutral sugars of the sheath fraction of *Gloeothece* PCC 6909 in thin layer chromatography. GlcA = Glucosamine; St = Standard sugar mixture: Gal, galactose; Glc, glucose; Man, mannose; Fuc, fucose; Xyl, xylose; Rib, ribose and Rha, rhamnose. 1 = Sheath fraction isolated after sucrose gradient centrifugation; 2 = Sheath fraction after lysozyme and hot SDS treatment; x_1 = unknown sugar.
Running solution: ethal acetate: pyridine: water = 12:5:4; running time: 2 h; colour developed with anilinpthalet.

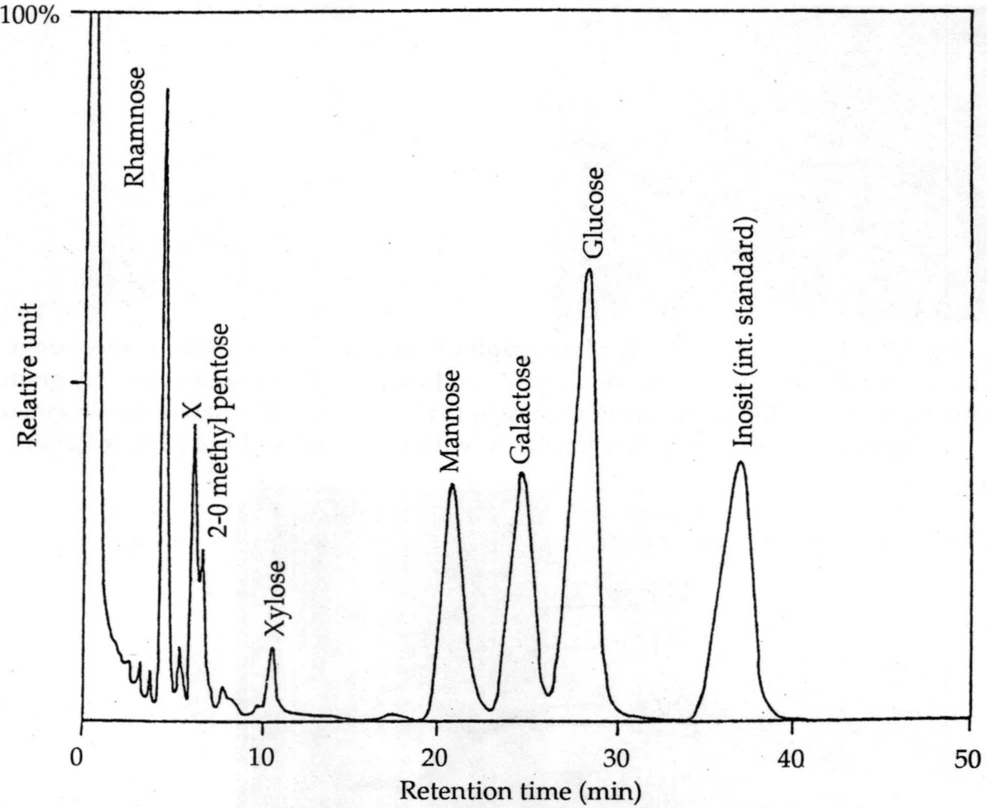

Fig. 24: Gas chromatographic analysis of neutral sugars of sheath fractions of *Gloeochece* **PCC 6909 isolated from sucrose gradient. Column: ECNSS-M; column temperature: 185° C isotherm; carrier gas: N$_2$ (30 ml/min).**

Table 15 : Chemical composition (% of dried material) of the isolated sheath fraction of *Gloeothece* PCC 6909.

Constituent	Sheath fraction After sucrose gradient separation	After lysozyme and SDS treatment
Sugar		
Unknown anhydro sugar X	0.9	2.4
Rhamnose	2.1	4.2
2-O-methyl pentose (not further identified)	0.5	0.8
Xylose	0.4	-

Constituent	Sheath fraction After sucrose gradient separation	After lysozyme and SDS treatment
Mannose	4.0	6.3
Galactose	4.6	7.5
Glucose	8.9	10.2
Total sugars	21.4	31.4
Protein	12.9	8.4
Glucosamine	2.4	-
Muramic acid	-	-
Diaminopimelic acid	-	-
Uronic acid	1.9	-
Phosphate	2.1	0.26
Total fatty acids	1.0	0.8

- = absent

Lysozyme: 5 mg in 25 ml ammonium acetate buffer, pH 6.8; 37° C, 12 h.

SDS: 4% (w/v) in Tris/HCl buffer, pH 7.8; 100° C, 15 min.

Only about 50% of the sheath fraction dry weight of the blue green algae so far reported has been identified. This may be due to the observed partial insolubility of the sheath under the hydrolytic conditions used and by a possible degradation of identified compounds during hydrolysis. Insolubility, together with mechanical stability and resistance to enzymatic treatment possibly render the sheath layers from blue green algae relatively stable against destruction in their natural environment.

Table 16 : Chemical composition (% of dried material) of the isolated sheath fractions (after lysozyme and SDS treatment) of *Gloeothece* PCC 6909 grown in presence of 10% (v/v) effluents (in BG 11 medium) of molasses distillery of Aska, fertilizer factory of Talcher and sponge-iron factory of Palasponga.

| Constituent | Sheath fraction of *Gloeothece* PCC 6909 Grown in BG 11 medium + 10% effluent of | | |
	Molasses distillery	Fertilizer factory	Sponge-iron factory
Sugar			
Unknown anhydro sugar	3.9	3.1	3.0

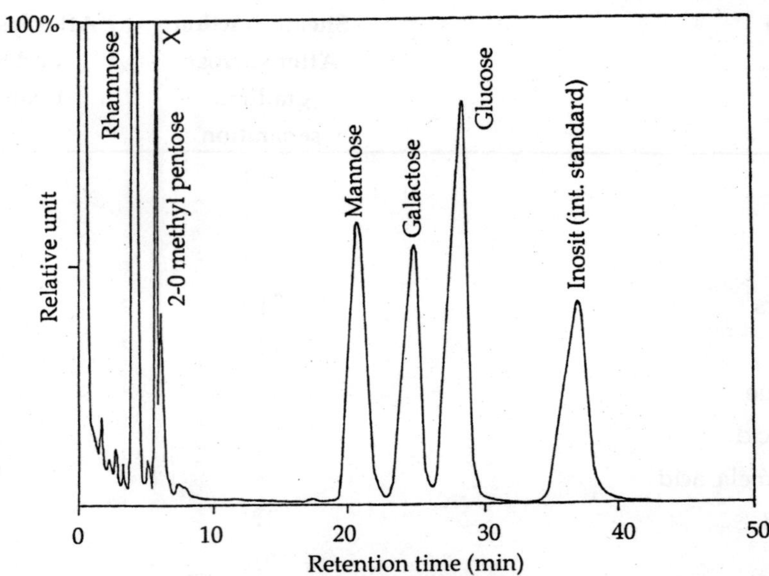

Fig. 25: Gas chromatographic analysis of neutral sugars of sheath fractions of *Gloeochece* **PCC 6909 isolated from sucrose gradient and purified with lysozyme and hot SDS treatment. Column: ECNSS-M; column temperature: 185° C isotherm; carrier gas: N$_2$ (30 ml/min).**

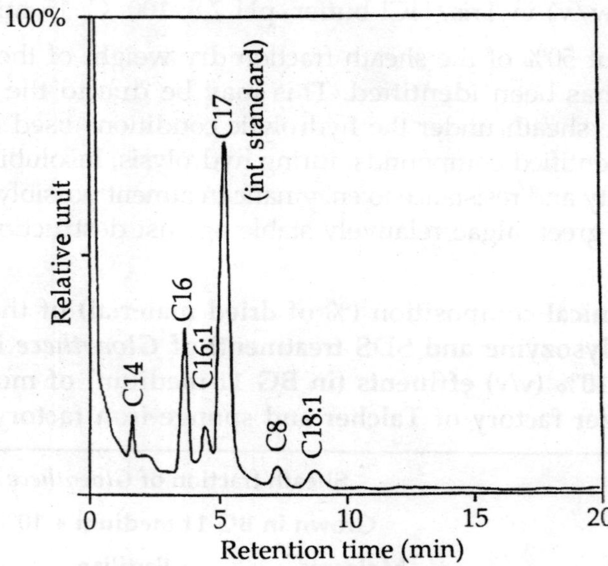

Fig.26: Gas chromatographic analysis of fatty acids of the sheath fractions of *Gloeothece* **PCC 6909 isolated from sucrose gradient and purified with lysozyme and hot SDS treatment. Column: EGSS-X; column temperature: 180° C, isotherm; carrier gas: N$_2$ (30 ml/min).**

Table-16 Contd. ...

| Constituent | Sheath fraction of *Gloeothece* PCC 6909 | | |
| | Grown in BG 11 medium + 10% effluent of | | |
	Molasses distillery	Fertilizer factory	Sponge-iron factory
Rhamnose	4.6	3.9	3.6
2-O-methyl pentose (not further) identified	1.0	1.3	0.9
Xylose	0.2	0.7	0.3
Mannose	7.8	6.2	6.4
Galactose	10.5	9.3	8.6
Glucose	14.3	11.8	10.5
Total sugars	42.3	36.3	33.3
Protein	6.2	13.6	10.4
Glucosamine	-	-	-
Muramic acid	-	-	-
Diaminopimelic acid	-	-	-
Uronic acid	1.8	1.3	2.1
Phosphate	1.2	0.39	1.4
Total fatty acids	0.34	0.52	0.67

Lysozyme: 5 mg in 25 ml ammonium acetate buffer, pH 6.8; 37° C, 12 h.

SDS: 4% (w/v) in Tris/HCl, pH 7.8; 100° C, 15 min.10% (v/v) proportions of the effluent of the industries were prepared in BG 11 medium (Rippka *et al.* 1979).

The sheaths of *Calothrix parietina* and *C. scopulorum* have been found to bind heavy metals, which were enriched in the order of Ni> Cu> Zn> Fe> Co> Mo, relative to their concentration in the medium (Weckesser *et al.*, 1988). Though cadmium is not an essential nutrient of blue green algae, its substantial adsorption to the sheaths of *Gloeothece* ATCC 27152 indicates the appreciable metal binding capacity of this extracelluar investment (Tease and Walker, 1987). The variation in composition observed in the sheath material collected from cultures grown with or without combined nitrogen sources resulted in differences in metal-binding capacity. A comparable concentration of metals has been described for the sheath of *Microcystis* from a natural habitat (Amemiya and Nakayama, 1984). Thus one of the functions of the external layers may be concentrating essential elements as well as toxic substances present at submarginal concentrations and also in excessive amounts at the cell surface enabling to proliferate in the environment otherwise uninhabitable for the organism. The preferential increase of sugar residues in the sheath of *Gloecothece* PCC 6909 when grown in presence of 10% (v/v) distillery effluent containing 4.5 to 6 per cent reducing sugars as observed in the present study indicate the utilization of sugars in specific biosynthetic process leading to their incorporation in the sheath layers. Further, the work reported earlier (Chapter 3.1.3.2) that formation of additional sheath layer of *Calothrix marchica* was observed in the medium supplemented with exogenous carbon compounds indicate the overall response of the organism to distillery effluent was determined by an individual adaptive response. This probably lead to the appearance of a new structural type of *Calothrix marchica* with elongated trichomes enveloped by thick sheath layers (Fig. 5), corresponding to the reorganised metabolic apparatus of the blue green alga in response to the changed environmental conditions.

❑❑❑

□ **4**

Response of Nitrogen fixing Blue Green Algae of Rice Fields to Pesticides

The application of nitrogen-fixing blue green algae in rice cropping as biofertilizer has gained much significance in rice cultivation in India (Venkataraman, 1981; Singh, 1985; Goyal, 1993). Reduction of chemical fertililzer input up to 30 per cent by supplementing with blue green algae is a significant finding when the conservation of energy is contemplated (Venkataraman, 1978; Roger and Kulasooriya, 1980). One of the important problems that has been noticed in mass multiplication tanks as well as under field conditions is the destruction of blue green algal populations by pesticide applications intended to control the insects and pests of the rice crop (Venkataraman, 1972; Kannaiyan, 1978). These agrochemicals also damage a wide variety of beneficial microorganism because of their persistence in the environment (Padhy, 1985). Therefore, pesticides used in routine applications in rice fields have important ecological effects in addition to those usually intended.

Several workers have reported the toxicity of a variety of herbicides, fungicides, insecticides and other pesticides to pure cultures of blue green algae (Table -2). The results of these investigations indicated that the effect of individual agrochemical on individual species is extremely variable, but generally harmful. Further, though there exists an inter-and intra-specific variation among blue green algae in their response to pesticide toxicity, generally recommended field dose of many of these agrochemicals did not affect their biological activity much, though excessive application had an adverse effect on nitrogen fixation thereby decreasing crop productivity.

A number of publications have suggested appropriate directions for future research efforts on the studies on blue green algae-pesticide interactions. They were: (i) definitive evaluation of the activity of blue green algae in agrochemical enriched soils and their importance to soil fertility (McCann and Cullimore, 1979), (ii) determinations of pesticide effect on blue green algae under field conditions (Stratton, 1987); (iii) evaluation of relative pesticide sensitivities of different blue green algal types (Wright, 1978; Pipe, 1987), and (iv) investigation of possible blue green algal resistance to pesticides and subsequent development of resistant, nitrogen-fixing blue green algal species/strains for their use as biofertilizer for rice cultivation (Padhy, 1985). Such studies can substantially improve the understanding of the complexities of blue green algal behaviour in pesticide-contaminated soils/blue green algae-pesticide interaction in the rice field ecosystem.

In general, laboratory studies are often conducted for determining pesticide-blue green algae interactions because the individual species can be isolated for study. Blue green algal species seem to be by far the most ubiquitous and important in aerobic waters of rice fields (Stewart, 1974). Since tests on the toxicity of pesticides to pure cultures of blue green algae showed a wide range of tolerance among the species investigated, it is, therefore, very difficult to assess the risk to fertility from pesticide use (Venkataraman, 1975). Thus field studies on the blue green algal population in pesticide burdened soils is required to be supplemented to data generated in the laboratory for proper analysis. Further, since all these organisms usually occur in association in the natural habitats, for understanding the pesticide effect on the photosynthetic microorganisms in the fields, interaction of pesticides also with other micro-algae viz. chlorophyceae and bacillariophyceae members, most commonly occuring in aquatic habitats need to be considered.

Apart from few published records (Venkataraman and Rajyalaxmi, 1972; Tarar and Shawale, 1984), the effect of pesticides on the soil algae of rice field soils has not been studied. In the present work effect of four commonly used pesticides, e.g. Furadan (carbofuran, 3% G), Sevin (carbaryl, 50%), Rogor (dimethoate, 30%) and Endotaf (endosulfan, 35%) belonging to carbamate, organophosphate and organochlorine groups, applied extensively to control rice pests of this region, on the occurrence of soil algae in the rice field soils was determined. Further, response of two heterocystous, nitrogen-fixing blue green algae isolated from the rice fields, e.g. *Westiellopsis prolifica* Janet and *Calothrix marchica* Lemm. possessing a thin and thick sheath layer around their trichomes respectively to these pesticides in laboratory culture was studied with a view to find out whether thick-sheathed blue green algal forms are relatively more tolerant to the toxic agrochemicals.

4.1 MATERIALS AND METHODS

Surface soil samples were collected in polythene bags from paddy fields of Keonjhar (Orissa) and its adjoining area. A composite soil sample was made by mixing the sample taken from different places. Algae present in the control soil and the pesticide treated soil samples were studied in Allen and Arnon's medium (1955) in presence or absence of combined nitrogen. Four commercial grade pesticides: two carbamate pesticides, Furadan and Sevin (Union Carbide Ltd.) and one each of organophosphate and organochlorine, Rogor and Endotaf respectively (Rallies India Ltd.) were used. Stock solutions of these pesticides were prepared freshly for experiments in the sterilized media and added to the culture media to obtain the desired concentration (100, 250, 500 and 1,000 ppm). The pH of all the media were adjusted to 7.5. Ten gram proportions of air dried soil was taken in cotton stoppered, sterilized 100 ml capacity conical flasks and 50 ml of the culture medium (with or without combined nitrogen and different concentrations of various pesticides) were added into each flask. After addition of nutrient solution to the culture, samples were agitated to ensure an uniform distribution of the pesticides. For each treatment, triplicates were set up and incubated at 28±2°C under 7.5 W/m² light intensity from fluorescent tubes. The algal forms appeared in the culture after 30 days of incubation were identified using standard monographs (Prescott, 1951; Smith, 1950; Desikachary, 1959; Hustedt, 1930). The survival percentage of the organisms were calculated on the basis of the number of species present in the respective treatments.

To study the response of nitrogen fixing blue green algae of rice fields to the pesticides in laboratory culture, the organisms, *Westiellopsis prolifica* Janet and *Calothrix marchica* Lemm. isolated from the soil samples of rice fields were used as the experimental organisms. The pesticides used and the culture conditions were the same as described previously while studying pesticides effect on soil algae. Experiments were conducted by inoculating equal amounts of actively growing organisms into 25 ml of medium in 100 ml cotton stoppered conical flasks. Growth was estimated on a dry weight basis and also in terms of absorbancy of acetone extracts in a Systronics spectrocolorimeter at 660 nm. Total nitrogen fixed by the blue green alga in presence of different concentrations of the pesticides was estimated on the 15th day of incubation. The cultures were subjected to Kjeldahal digestion and total nitrogen was estimated as per Herbert *et al.* (1971). Corrections were made for the nitrogen present in the pesticides.

Survivability of the organisms was studied on agar plates containing varying concentrations (10 to 1,500 ppm) of the pesticides. Samples of 1.0 ml of a diluted suspension of the blue green alga was spread aseptically onto each of the agar plates and inoculated under light in the culture room. After 10 days of incubation, the colonies were counted under a binocular microscope and the survival data were plotted taking survival on the control plates as 100%. The rate of O_2 evolution (photosynthesis) was measured with a Clark type oxygen electrode (Yellow Spring, Ohio, USA) and a Hitachi 056 recorder. Cyanobacterial filaments containing approximately 2.0 to 3.0 µg of chlorophyll-*a* from the 72 h-old experimental culture, containing various concentrations of the pesticides were transferred to a periglass vessel fitted with an outer Jacket for water circulation and temperature maintenance (25°C) during the experiment. To avoid reduction at the cathode and to maintain a stable oxygen gradient across the membrane, the cell suspension was stirred with a magnetic stirrer. Saturating light intensity of 200 W/m² was provided by a 150 W halogen lamp through a heat cut filter (70%). The signal from the cathode was fed through a preamplifier to increase the efficiency of recording. Chlorophyll-*a* was extracted from the blue green algal filaments with 80% (v/v) acetone and the amount was determined spectrophotometrically using the absorption coefficient of Mackinney (1941).

4.2 RESULTS AND DISCUSSION

4.2.1 Effect of pesticides on algae of rice field soils

Twenty seven algal taxa appeared in the composite rice-field soils in the control flasks, of which 17 belonged to cyanophyceae, 4 to chlorophyceae and 6 to bacillariophyceae (Table-17). Of the cyanophyceae members, 10 species were heterocystous, which have also been identified from the rice-field soils of other regions of India (Singh, 1961; Venkataraman, 1969; Sahu   *et al.*, 1996; Nayak *et al.*, 1996). *Nostoc commune, Aulosira ferilissima* and *Westiellopsis prolifica*, which were present abundantly in the test soils, have been reported to fix atmospheric nitrogen in pure culture (Fogg *et al.*, 1973; Nayak *et al.*, 1996), thus suggesting the richness of beneficial blue green algae in the rice fields soils of this region. The occurrence of microalgae were remained almost unaffected even in presence of Furadan and Sevin at 100 ppm dose in the soil. However, their quantitative occurrence was decreased to a considerable extent. With increase in the dose level, proportionate decrease in their occurrence was noted. Bacillariophyceae members disappeared first and then the species belonged to chlorophyceae in presence of more than 100 ppm of Furadan and Sevin. At 500 ppm dose level of the carbamate pesticides most of the unicellular and filamentous heterocystous forms of cyanophyceae and the members of chlorophyceae and bacillariophyceae could not grow in the soils. Except the blue

Table 17 : Qualitative occurrence of soil algae at the end of 30 days after treatment of various pesticides

		Furadan (ppm)				Sevin (ppm)				Rogor (ppm)				Endotaf (ppm)			
		100	250	500	1000	100	250	500	1000	100	250	500	1000	100	250	500	1000
Chroococcus minutus (Kuz.) Nag	++	+	+	+	-	+	+	+	-	+	+	-	-	+	-	-	-
Gloeocabsa atrata (Turp.) Kutz		+	+	+	+	+	+	+	+	+	+	-	-	+	+	-	-
Aphanothece naegelel Wart	++	+	+	-	-	+	+	-	-	+	-	-	-	-	-	-	-
Oscillatoria tenuis (Ag.) Gom.	++	+	+	+	-	+	+	+	-	+	+	-	-	+	-	-	-
Lyngbya major (mene) Gom.	++	+	+	+	+	+	+	+	+	+	+	+	-	+	+	+	-
Anabaenopsis circuiaris (West M)		+	-	-	-	+	+	-	-	+	+	-	-	+	-	-	-
Cynorospermum muscicola(Born et Flah)		+	+	-	-	-	+	-	-	+	-	-	-	-	-	-	-
Nostoc punctiforme (Kutz. Hariot)		+	+	-	-	+	+			+	+	-	-	-	-	-	-
Nostoc commune (Vau Born et Flah)	++	+	+	+	-	+	+	+	-	+	+	-	-	+	-	-	-
Nostoc piscinale (Kutz)		+	+	-	-	+	+	-	-	+	+	-	-	+	-	-	-

	Furadan (ppm)				Sevin (ppm)				Rogor (ppm)				Endotaf (ppm)			
	100	250	500	1000	100	250	500	1000	100	250	500	1000	100	250	500	1000
Anabaena oryzae Fritsch	+	+	-	-	+	-	-	-	+	-	-	-	-	-	-	-
Aulosra fertilissima Ghose ++	+	+	+	-	+	+	-	-	+	+	-	-	+	-	-	-
Scytonama pascheri Bhardwaj	+	+	+	+	+	+	+	-	+	+	-	-	+	-	-	-
Calothrix marchiea Lemm.	+	+	+	+	+	+	+		+	+	+	-	+	+	-	-
Haplosiphon welwitschii West	+	-	-	-	-	-	-	-	+	-	-	-	-	-	-	-
Westiellopsis prolifica Janet (++)	+	+	-	-	+	+	+	-	+	+	-	-	+	-	-	-
Scenedesmus obllquus Trup. (++)	+	+	-	-	+	+	+	-	+	+	-	-	+	-	-	-
Ulothrix zonata (Web. et. Mohr) Kutz.	-	-	-	-	+	-	-	-	-	-	-	-	-	-	-	-
Oedogonium oblogollum Kirch. (++)	+	+	-	-	+	-	-	-	+	-	-	-	+	-	-	-
Cosmarium portianum Arch.	+	-	-	-	+	+	-	-	-	-	-	-	-	-	-	-
Fragilana pinnnata Her. (++)	+	-	-	-	-	-	-	-	-	-	-	-	-	-	-	-
Navicula pupula Kutz.	+	+	-	-	+	-	-	-	-	-	-	-	-	-	-	-

		Furadan (ppm)				Sevin (ppm)				Rogor (ppm)				Endotaf (ppm)			
		100	250	500	1000	100	250	500	1000	100	250	500	1000	100	250	500	1000
NavIcula mimma Grun	(++)	+	-	-	-	-	-	-	-	-	-	-	-	-	-	-	-
Cyclotella menenghiniana Kutz.		+	-	-	-	+	-	-	-	-	-	-	-	-	-	-	-
Gomphonema lanceolatus Ehr.		+	-	-	-	-	-	-	-	-	-	-	-	-	-	-	-
Nitzchia polaris Grun.	(++)	+	+	-	-	+	-	-	-	-	-	-	-	-	-	-	-
Total number of algal species appeared (Control soil = 27)		26	19	8	4	24	16	9	2	19	13	3	0	14	4	0	0
% survival (control = 100)		96.2	70.3	29	14.8	88.8	59.2	33.3	7.4	70.3	48	11.1	0	51.8	14.8	0	0

- = absent ; + = present ; ++ = abundant

green algal species, e.g. *Gloeocapsa atrata, Lyngbya major, Calothrix marchica* and *Scytonema pascheri*, which possess well defined sheath around their cells/ trichomes, none of the other taxa could tolerate 1,000 ppm of Sevin and Furadan.

The effect of the organophosphate and organochlorine pesticides, Rogor and Endotaf respectively, on the occurrence of bacillariophyceae and chlorophyceae forms were quite specific. These organisms were almost completely eliminated from the soils in presence of more than 100 ppm of the pesticides. Only *Scenedesmus obliquus* and *Oedogonium obliongollum* survived in presence of 100 ppm of these pesticides. With further increase in the pesticide concentrations, almost all the heterocystous forms of cyanophyceae and the chlorophyceae and bacillariophyceae members did not grow in the cultures. Only *Lyngbya major* and *Calothrix marchica* tolerated up to 500 ppm of Rogor. However, few filaments of *Lyngbya major* appeared in the soils supplemented with 500 ppm of Endotaf after approximately 40 days of incubation. An analysis of these results, on the basis of total number of different groups of algae appeared in the culture and their survival percentage in presence of various concentrations of the pesticides (Table-18) indicate that rice-field soil algae show variable resistance to pesticide treatment. The organochlorine pesticide, Endotaf was more toxic than the organophate, Rogor and the carbamates, Sevin and Furadan in sequence. However, higher concentrations of all these pesticides greatly affect their survival showing high algicidal potential. In general, sensitiveness of different group of organisms to pesticide application was found to be in the order of: bacillariophyceae > chlorophyceae > sheathless heterocystous and unicellular cyanophyceae members> heterocystous ensheathed and non-heterocystous ensheathed forms. The mechanism of resistance of the ensheathed blue green algae to the pesticides is possibly due to the presence of sheath outside the cell wall of these organisms which might be protecting them from the adverse effect of the toxic chemicals. From the results obtained in the present investigation it was also seen that with higher dose of pesticide application in the rice-fields, i.e. more than 100 ppm of Furadan, Sevin and Rogor and even 100 ppm of Endotaf, qualitative and quantitative occurrence of heterocystous blue green algae decreased considerably. This suggests that indiscriminate use of these pesticides may cause adverse effects on the nitrogen-fixing blue green algae of rice fields, which has a direct influence on total productivity.

4.2.2 Growth response, photosynthetic oxygen evolution and nitrogen fixation of two different species of rice field blue green algae to pesticides in laboratory culture

The effect four different pesticides on the survival; growth and nitrogen fixation of *Westiellopsis prolifica* and *Calothrix marchica* was studied in experiments with 10, 20, 50, 100, 250, 500, 1000 and 1500 ppm of each pesticides in the culture solution. The results, Table-19 to 21, indicate a progressive decrease in the growth and nitrogen-fixation of both the alga with increasing concentrations of the pesticides; however, *C. marchica* was always more tolerant than *W. prolifica* to increasing pesticide doses. In laboratory culture also the carbamate pesticides, Furadan and Sevin were less toxic than the organophosphate, Roger and the organochlorine, Endotaf to blue green algae. The presence of 1,000 and 3,000 ppm of Furadan and Sevin in the medium were algistatic for *W. prolifica* and *C. marchica* respectively. But similar respective concentration of Rogor and still a lower concentration of Endotaf were lethal for both the algae (Fig. 27). With Roger, the growth of *W. prolifica* remained almost unaffected at 10 ppm, but at the same concentration, the carbamate insecticides stimulated the growth of the organism. Nitrogen fixation of *W. prolifica* also increased at these lower doses (Table-21). To the contrary, *C. marchica* which possess a well defined sheath layer around the trichome, showed higher growth and nitrogen fixation over control in presence of up to 50 ppm of Furadan, 20 ppm of Sevin and 10 ppm of Roger. Similar enhancement of growth and nitrogen fixation in various other blue green algal species in presence of lower doses of Furadan (Kar and Singh, 1978), 2,4 D (Das and Singh, 1977) and Sevin (Adhikary *et al.*, 1984) have been reported. At concentrations above 20 ppm and 50 ppm of the carbamate pesticides for *W. prolifica* and *C. marchica* respectively, and 10 ppm of Rogor for both the species, growth and nitrogen fixation progressively declined (Table - 19 to 21). Endotaf was toxic at 10 ppm for *C. marchica* and *W. prolifica*. Growth and nitrogen fixation of the former and latter species was ceased in presence of 500 ppm and 250 ppm of the insecticide respectively.

The effect of pesticides on photosynthetic oxygen evolution of both the blue green algae is shown in Table-21. Incubation cf the organisms for a period of 72 h in the presence of varying concentrations of pesticides and determination of the rate of photosynthesis there after (in terms of O_2 evolved/mg chlorophyll *a*/h) showed that this process was inhibited by all the tested pesticides. However, lower concentrations of the carbamate pesticides (up to 10 ppm) did not show any adverse effect on the photosynthesis of *C. marchica*. The commercial grade Furadan was less toxic to both the organisms than Sevin, and further, decrease of rate of photosynthesis of *C. marchica* was less than that of *W. prolifica* upon exposure to similar dose of the insecticides (Table - 22). Oxygen evolution of *W. prolifica* was completely suppressed on exposure to over 500 ppm of Rogor. The toxic effect of Endotaf on the photosynthetic oxygen evolution of the organism was more pronounced even at half the dose of Rogor. However, *C. marchica* continued to photosynthesize even in presence of up to 500 ppm of Endotaf and 1000 ppm of Rogor, though further increase in their concentrations in the culture medium ceased the Oxygen evolution capacity of the organism.

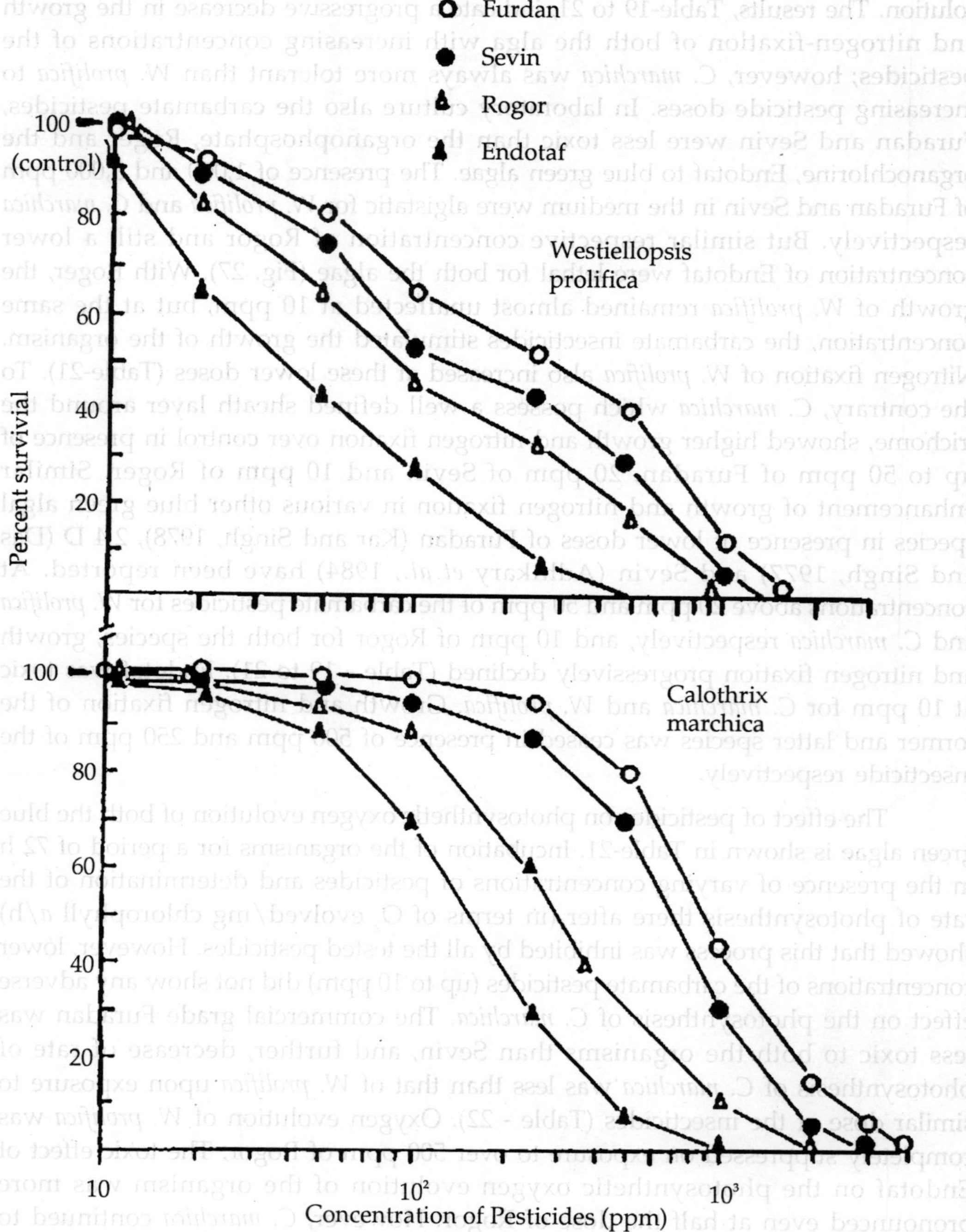

Fig. 27: Per cent survival of *Westiellopsis prolifica* and *Calothrix marchica* in the presence of different concentrations of pesticide in the medium. Cultures were incubated at 28±2°C under 7.5 W/m² light intensity up to 15 days.

Table 18 : Total number of different groups of algae appeared in the culture and their survival percentage in presence of various concentrations of pesticides

Pesticide	Concentration (ppm)	Cyanophyceae (Blue green algae)				Chlorophyceae		Bacillariophyceae	
		Heterocystous		Non-heterocystous					
		Total no of species	% Survival	Total no of species	% Survival	Total no. of species	% Survival	Total no. of species	% Survival
Control	Without pesticide	12		5		4		6	
Furadan	100	12	100	5	100	4	100	5	83.3
	250	10	83.3	5	100	2	50	2	33.3
	500	4	33.3	4	80	0	0	0	0
	1000	2	16.6	2	40	0	0	0	0
Sevin	100	12	100	5	100	4	100	2	33.3
	250	7	58.3	5	100	2	50	0	0
	500	4	33.3	4	80	1	25	0	0
	1000	0	0	2	40	0	0	0	0
Rogor	100	12	100	5	100	2	50	0	0
	250	8	66.6	4	40	1	25	0	0
	500	2	16.6	1	20	0	0	0	0
	1000	0	0	0	0	0	0	0	0
Endotaf	100	8	66.6	4	80	2	50	0	0
	250	2	16.6	2	40	0	0	0	0
	500	0	0	1	20	0	0	0	0
	1000	0	0	0	0	0	0	0	0

Table 19 : Effect of different concentrations of pesticides on the growth (dry weight in mg) of *Westiellopsis prollifica* and *Calothrix marchica*. Harvested after 15 days of incubation

Concentration of pesticide (ppm)/ 25 ml culture	Furadan		Sevin		Rogor		Endotaf	
	W. prolifica	*C. marchica*	*W. prolifica*	*C. marchica*	*W. prolifica*	*C. marchica*	*W. prolifica*	*C. marchica*
Control (without pesticide	28.5 ±1.3	25.7±1.6	28.5± 1.9	25.7±2.1	28.5± 1.6	25.7±1.9	28.5±2.3	25.7± 2.2
10	30.0± 1.8	28.1± 2.3	30.2 ± 2.2	29.4 ± 0.4	28.9± 1.5	25.9±2.1	26.0±1.4	24.3±1.7
20	32.8 ± 2.2	33.3 ± 3.1	26.9 ± 1.3	29.7 ±1.9	25.4 ±1.9	24.4±2.8	20.2±1 .1	21.6±1.1
50	27.2 ± 1.7	30.4 ± 3.6	22.8 ± 2.1	21.5±2.2	18.6±0.7	20.9±1.3	14.5±!:2.6	18.9:t2.6
100	22.4 ± 1.1	24.9 ± 1:2.7	16.2 ± 1.6	19.1±1.1	12.8 ±1.2	13.7±1.4	6.9±0.8	11.4±1.3
250	15.7 ± 1.3	23.2 ± 1.6	12.5 ± 0.9	16.6±1.3	9.2 ±0.9	9.1±0.8	2.8±0.6	6.8±0.7
500	6.4 ± 0.7	18.5±1.9	5.6 ± 0.4	12.4±1.2	3.8±0.4	5.4±0.5	1.0±0.2	3.3±0.8
1000	1.8 ± 0.3	11.7 ± 2.3	1.0 ± 0.3	8.1 ±0.7	0.8±0.1	2.8±0.7	0.6±0.04	1.0±0.09
2000	0.5 ± 0.04	4.4 ± 0.8	0.4 ± 0.06	2.6±0.2		0.6±0.07		0.4±0.06
3000		1.9 ± 0.7		0.7±0.08				
4000		0.3± 0.04						

Dry weight of the initial inoculum = 1.0 mg; values represent mean of three replicates ± S.D.

Table 20 : Effect of different concentrations of pesticides on the growth (Cabsorbance of acetone dissolved pigments at 660 nm/25ml culture) of *Westiellopsis prollifica* and *Calothrix marchica*. Harvested after 15 days of incubation

Concentration of pesticide (ppm)/ 25 ml culture	Fkuradan		Sevin		Rogor		Endotaf	
	W. prolifica	*C. marchica*	*W. prolifica*	*C. marchica*	*W. prolifica*	*C. marchica*	*W. prolifica*	*C. marchica*
Control (without pesticide	0.34	0.27	0.34	0.27	0.34	0.27	0.34	0.27
10	0.36 (±5.88)	0.3 (+11.1)	0.35(+2.9)	0.29(+7.4)	0.34 (±0)	0.28 (+3.7)	0.03(-11.7)	0.25(-7.4)
20	0.34 (±0)	0.32 (+18.5)	0.3 (-11.7)	0.29(+7.4)	0.25(-26.5)	0.22(-18.5)	0.21(-38.2)	0.21(-22.2)
50	0.28 (-17.6)	0.28 (+3.7)	0.26 (-23-5)	0.25 (-7.4)	0.18(-47.0))	0.18 (-33.3)	0.09(-73.5)	0.16(-40.7)
100	0.23(-32.3)	0.22(-18.5)	0.2(-41.2)	0.21 (-25.1)	0.11(-57.6)	0.15(-44.4)	00.03(-91.3)	0.1 (-62.9)
250	0.18 (-47.0)	0.17 (-37.0)	0.15 (-55.8)	0.14 (-48.1)	0.04 (-88.2)	0.11(-54.2)	0.01(-97.0)	0.05(-81.5)
500	0.08 (-76.5)	0.12 (-55.5)	0.06 (-82.3)	0.11 (-59.2)	0.015(-95.6)	0.09(-66.6)	0	0.02(-92.6)
1000	0	0.09(-66.6)	0	0.07 (-74.0)	0	0.05(-81.5)		0
2000		0.05 (-81.5)		0.03 (-88.9)		0		
3000		0		0				
4000								

Absorbance of acetone dissolved pigments at 660 nm of the initial inoculum = 0.02; values represent mean of three replicates; figures in parenthesis show percent increase (+) or decrease (–) over control.

Table 21 : Effect of different concentrations of pesticides on the total nitrogen fixed (mg/25 ml culture) by *Westiellopsis prolifica* and *Calothrix marchica*. Hravested after 15 days of incubation

Concentration of pesticide (ppm)/ 25 ml culture	Fkuradan		Sevin		Rogor		Endotaf	
	W. prolifica	*C. marchica*	*W.prolifica*	*C. marchica*	*W. prolifica*	*C. marchica*	*W. prolifica*	*C. marchica*
Control (without pesticide	3.6	3.1	3.6	3.1	3.6	3.1	3.6	3.1
10	3.8(+5.5)	3.4(+9.6)	3.5(+2.7)	3.3(+6.4)	3.8(+5.5)	3.3(+6.4)	1.7(-52.7)	2.7(-12.9)
20	3.4(-5.5)	3.7(+19.3)	3.2(-11.1)	3.5+12.9)	2.7(-25.0)	2.8(- 9.7)	1.1(-69.4)	2.1(-32.2)
50	2.8(-22.2)	3.3(+6.4)	2.2(-38.8)	29(-6.4)	2.0(-44.4)	2 1(-32.2)	0.83(-79.6)	1.4(-54.8)
100	2.0 (-44.4)	2.8(-9.7)	1.5(-58.3)	1.7(-45.1)	0.9(-74.7)	1.5(- 51.6)	0.25(- 93.0)	0.82(-73.5)
250	1.6(-55.5)	2.3(-25.8)	0.96(-73.3)	1.0(-67.7)	0.33(-90.8)	0.92(-70.3)	0	0.24(-92.2)
500	0.83(-76.9)	1.7(-45.1)	0.45(-87.5)	0.64(-79.3)	0.1(-97. 2)	0.47(- 84.8)		0
1000	0	0.93(-70.0)	0	0.35(-88.3)	0	0		
2000		0.43(-86.1)		0				
3000		0						
4000								

Values represent mean of three replicates; figures in parenthesis show per cent increase (+) or decrease (–) relative to the nitrogen fixation in the control.

Table 22 : Effect of different concentrations of pesticides on the photosynthesis in *Westiellopsis prolifica* and *Calothrix marchica*.

Concentration of pesticide (ppm)/ 25 m culture	μmol of O$_2$ evolved mg chl $^{a-1}$. h^{-1}							
	Furadan		Sevin		Rogo		Endotaf	
	W. prolifica	*C. marchica*	*W.porlifica*	*C. marchica*	*W. prolifica*	*C. marchico*	*W prolifiea*	*C. marehica*
C (without pesticide)	186	144	186	144	186	144	186	144
10	182(2.15)	144(0)	176(5.3)	144(0)	170(80.6)	142(1.4)	144(22.5)	136(5.5)
50	154(17.2)	141(2.1)	138(25.8)	138(4.2)	124(33.3)	121(15.9)	98.5 (47.0)	111 (22.9)
100	112(39.7)	139(3.4)	108(41.9)	124 (13.8)	96.4(45.1)	54{41.6)	33.2(82.1)	63(56.2)
250	76.6(58.8)	131(9.0)	72.4(61.0)	116 (19.4)	48.3(74.0)	.49(65.9)	12.2(93.4)	25(82.6)
500	44.1(76.1)	123(14.5)	22.2(88.0)	74(48.6)	7.4(96.1)	16(88.8)	0 (100)	2.8(98.0)
1000	7.8(95.8)	69(52.0)	2.3(98.7)	19(85.8)	0 (100)	7. 4(94.8)		0 (100)

The organisms were incubated in the presence of pesticides for a period of 72 h then the rate of photosynthesis was determined. Values represent mean of three replicates; figures in parenthsis show percent inhibition of photosynthetic oxygen evolution in the presence of the pesticides.

These results accord with the changes in the morphology of *W. prolifica* as well as of *C. marchica* exposed to pesticides. In presence of more than 250 ppm of Furadan, Sevin and Rogor and 100 ppm of Endotaf, the erect branches of the filaments of *W. prolifica* were suppressed and the number of heterocysts per filament decreased to a larger extent. *In C. marchica*, the length of the individual filament of the organism gradually reduced in size with increase of the dose of the pesticides, e.g. more than 500 ppm of Furadan and Sevin, 250 ppm of Rogor and 50 ppm of Endotaf in the culture medium. With further increase in the concentration of the respective pesticides, the cells in the trichomes became deformed, disorganized together with loss of chlorophyll. However, the sheath around the trichomes remained intact, though reduced to a thin layer even in presence of sublethal dose of the pesticides. At the lethal doses, the filaments of both the organisms bleached and the cells disintegrated in the culture. These results showed that higher concentrations of the pesticides decreased the survivability, growth and nitrogen fixation of rice field blue green algae. The EC 50 values of the pesticides on the growth of both the organisms shown in Fig. 28 clearly showed that *C. marchica* possessing a distinct sheath layer around its trichome tolerated comparatively higher doses of the agrochemicals than the thin sheathed *W. prolifica*. The results of chapter 4.2.1 also showed that the blue green algal species occurring in soil possessing a sheath layer tolerated higher concentrations of the pesticides than the sheathless forms.

The pesticide application rates recommended to control rice pests of this region are 0.75 and 1.0 kg ha^{-1} for carbaryl and carbofuran (Sevin and Furadan), and 0.5 and 0.8 litre ha^{-1} for endosulfan (Endotaf) and dimethoate (Rogor), respectively which will provide a range of 10-20 ppm in the rice fields. From the results obtained in the experiments on the effect of pesticides on soil algae, it was seen that higher levels of pesticide application, i.e. more than 100 ppm of Furadan and Sevin, and even 10 ppm of Rogor and Endotaf, adversely affected the occurrence of heterocystous forms of blue green algae which are responsible for nitrogen fixation in the rice fields. The toxic dose of organophosphate and carbamate pesticides in the experiments in laboratory culture of two different heterocystous blue green algae was still lower, e.g. more than 10 ppm for *W. prolifica* and a little higher concentration for *C. marchica*. This shows that though lower concentrations, < 10 ppm may not affect the growth and nitrogen fixation of blue green algae under field conditions, even recommended dose of the pesticides under waterlogged conditions in the rice fields decreased the growth and nitrogen fixation of blue green algae considerably. The organochlorine pesticide was especially more toxic; at more than 10 ppm affected the occurrence of blue green algae in the field as well as their growth in the laboratory culture. This shows that the insecticides used indiscriminately might adversely affect the survivability of a variety of microorganisms in nature. Since nitrogen fixing blue green algae are now used as

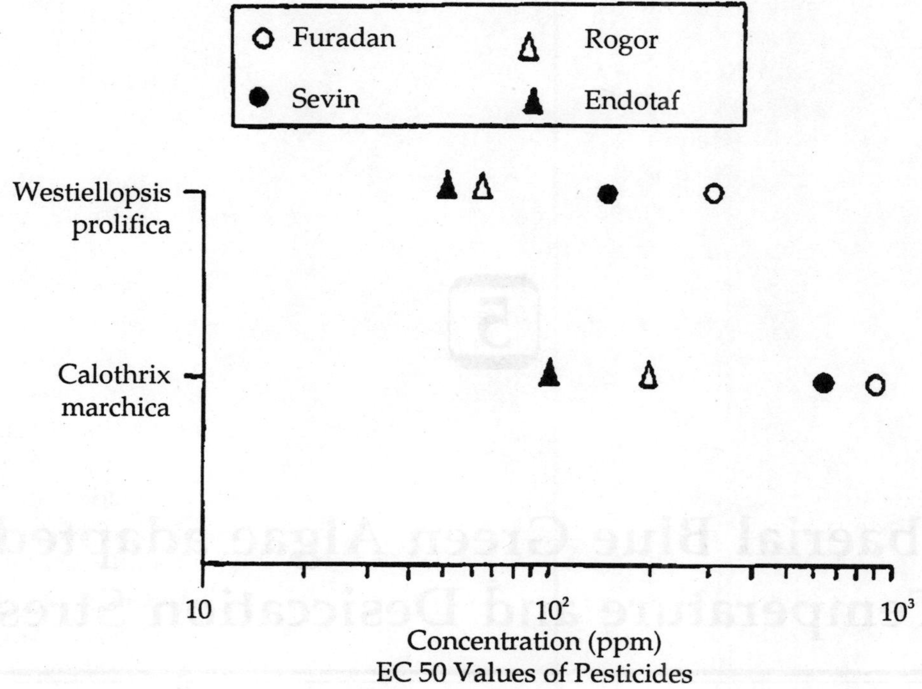

Fig. 28: Range of EC 50 values of the pesticides Furadan, Sevin, Rogor and Endotaf for growth of sheathless *Westiellopsis prolifica* and ensheathed *Calothrix marchica*.

biofertilizer for rice cultivation, it is essential to screen efficient strains those are capable of growing and fixing nitrogen at a higher rate even in presence of recommended doses of the agrochemicals including pesticides, before their inoculation in to the field. The blue green algal species possessing a well defined sheath layer outside their cell wall are such promising strains which can thrive in the pesticide burdened soils and also continue to grow and fix nitrogen in presence of relatively higher pesticide doses than the sheathless strains, and thus can be included in the multi-strain mixture for using as biofertilizer. Further, as pesticides affect adversely on the growth and nitrogen fixation of blue green algae at higher doses irrespective of the morphological specialization of these organisms, caution should be taken to determine the appropriate application dosage of these environmental pollutants before applying them into the rice fields.

□□□

5

Subaerial Blue Green Algae adapted to Temperature and Desiccation Stress

Certain blue green algae posses the ability to occur in a wide variety of extremely dry habitats like tree barks (Cox and Hightower, 1972), cave rocks (Friedmann, 1962), arid desert rocks (Potts *et al.*, 1953; Büdel and Wessels, 1991) and on surfaces of buildings coated with cement , lime or structures covered with asbestos cement sheets in the tropics (Tripathy and Talpasayi, 1980). Under such conditions they must therefore possess efficient mechanisms to prevent or counteract the effect of desiccation, elevated temperature and of strong solar radiation including those caused by the particularly deleterious UV-region of the spectrum. These environmental stresses are likely to play a determinative role in the ecophysiology of the organims that grow in aerophytic environments; on or inside of rocks, on and in soils and sediments. Blue green algal communities are known from several of such environments, e.g. deserts of south-west United States of America and Mexico (Fridmann, 1972); Corolado plateau in Arizona (Bell *et al.*, 1986); Savanna grassland of northern Transvall and Clarens sandstones of South Africa (Wessel and Büdel, 1995); Uluru rock of central Australia (Büdel and Wessels, 1991); Viscaino region and Neveda test sites (Shields and Drouet, 1962; Ocampo-Friedmann, 1976; Friedmann, 1977, 1982). In addition, on the exposed surface of monuments, archaeological remains and works of art, blue green algae are flourishing where sunlight and moisture are available. Blue green algal forms have been reported from cave walls and churches of Europe (Dupuy *et al.*, 1976; Giacobin *et al.*, 1979; Thomaselli *et al.*, 1979; Vander Molen *et al.*, 1980; Grilli-Caiola *et al.*, 1987; Albertano and Grilli-Caiola, 1989; Ortega-Calvo *et al.*, 1993), stone monuments of Rome and Athens (Anagnostidis *et al.*, 1983; Pietrini *et al.*, 1985) and on the exposed rock surface of temples and monuments (Adhikary and Satapathy, 1996). These organims impart blackish-brown coloration to the substratum due to their excessive growth forming characteristic crust or tuft layers.

The blue green algal species on such harsh habitats are the organisms with specialized adaptations for surviving under extreme conditions of temperature and desiccation. Such adaptations can either be physiological or morphological. These include the presence of a mucilage sheath and pigments in these sheaths to absorb moisture as well as to cut down excessive loss of moisture. Though resistance of several species of plants towards extreme water stress have been previously reported (Potts, 1994; Bewley, 1979), the precise mechanism of tolerance of subaerial blue green algae exposed to extreme heat and dryness is not fully understood

5.1 EPILITHIC SUBAERIAL BLUE GREEN ALGAE OCCURRING ON THE TEMPLE WALLS

Most of the temples and monuments of India built during 6th to 16th century A.D. using sandstone or khondalite rocks look blackish brown due to colonization of certain blue green algae. Excessive growth of these blue green algae in subsequent years provide suitable habitats to support appearance of bryophytes and other higher plants which greatly damage these monuments (Singh, 1993). Nothing is known about the organisms inhabiting the exposed rock surface of temples and other ancient monuments of India and nearby countries. During mid days of summer months (March-May) the temperature on the rock surface of temples goes beyond 60° C, coupled with high light intensity and extreme dryness. In such inhospitable environment, certain species of blue green survive forming characteristic vegetation as blackish-brown crust or tuft and resume their growth on the onset of monsoon (from mid June). Survey of these terrestrial blue green algal forms occurring on the rock surface of 32 different gigantic temples and ancient monuments at various locations of Orissa, Tamilnadu, Karnatak and Delhi states of India and of Nepal has been made.

The temple city of India, Bhubaneswar (in Orissa state) abounds with hundreds of temples of cultural heritage. In addition the famous Jagannath temple of Puri of Hindu faith and the Sun temple at Konark - "The World Heritage Monument" are located within 60 km radius area the so called "Golden Triangle". All these temples are built up of blocks of sandstone and/or khandalite of Gandawana age between 7th to 13th century A.D. Visual observations showed that different stone blocks on a particular face of a monument, exposed to similar environmental conditions, are colonized by blackish brown crusts comprised of a single species of blue green algae belonged to the genus *Tolypothrix*. Underneath the crusts, the stones are found to be withered to varying degrees. Detail observations on this species were made using both field samples and cultured specimens.

The blue green algae survived at the elevated temperature and extreme dryness for about four months (March-June) forming a dark brownish coloured crust on the rock surfaces. The organism possess a thick brownish sheath layer and absorb strongly in the near-UV-blue region. The colouration was possibly due to the presence of the brown sheath pigment, scytonemin absorbing in the UV region of the spectrum, which has been identified in many species of blue green algae from cultures and natural population exposed to intense solar radiation (Garcia-Pichel and Castenholz, 1991, 1993; Garcia-Pichel *et al.*, 1992, 1993; Proteau *et al.*, 1993). Presence of a UV-A/B absorbing pigment with maxima at 312 and 330 nm has also been described in a terrestrial blue green alga, *Nostoc commune* (Scherer *et al.*, 1988). These UV-absorbing pigments reported in a variety of desiccation tolerant blue green algae have been attributed to play a considerable role in photoprotection (Whitton, 1987, 1992). Thus absorption characteristics of methanolic extracts of this terrestrial blue green algae occurring predominantly on the rock surface of several temples of Orissa state and also of its isolated sheath fractions was studied.

5.1.1 Materials and methods

5.1.1.1 Study sites

Thirty two ancient temples and monuments built during 6th to 16th century A.D., located at different places of India have been chosen as the study sites (Fig. 29). Out of these, 14 temples e.g. Lingaraj, Kedar Gouri, Raja Rani, Bhaskareswar, Parsurameswar, Brahmeswar, Megheswar, Mukteswar, Rameswar and Khandagiri-Udayagiri cave temples of Bhubaneswar; Jogini temple, Hirapur; Madhav temple, Niali; Sun temple, Konark and Jagannath temple, Puri are in the coastal region of Orissa (19° 80' to 20° 24' N, 85° 87' to 86° 9' E); Nrusinghanath temple, Paikamal in the Western Orissa (20° 68' N, 82° 46' E); The temples e.g. Arjun's Pennance and Rath temples, Mahabalipuram; Kailashnath temple, Kancheepuram; Natraj temple, Chidambaram;Vrihadeswar temple, Tanzavur; Rock Fort and Sri Rangaswami temples, Tiruchirappalli and Meenakshi temple, Madurai are in Tamil Nadu state (9° 94' to 12° 94' N, 78° 98' to 80° 15' E); 6 temples, e.g. Hoysaleswara temple, Helebid; Channekeshwara temple, Belur; Gomateswar temple, Sravanabelagola; Bhutanath cave temple, Badami; Ladkhan temple, Aihole and Virupaksha temple, Pattadakal in Karnatak state (12° 64' to 15° 93' N, 75°

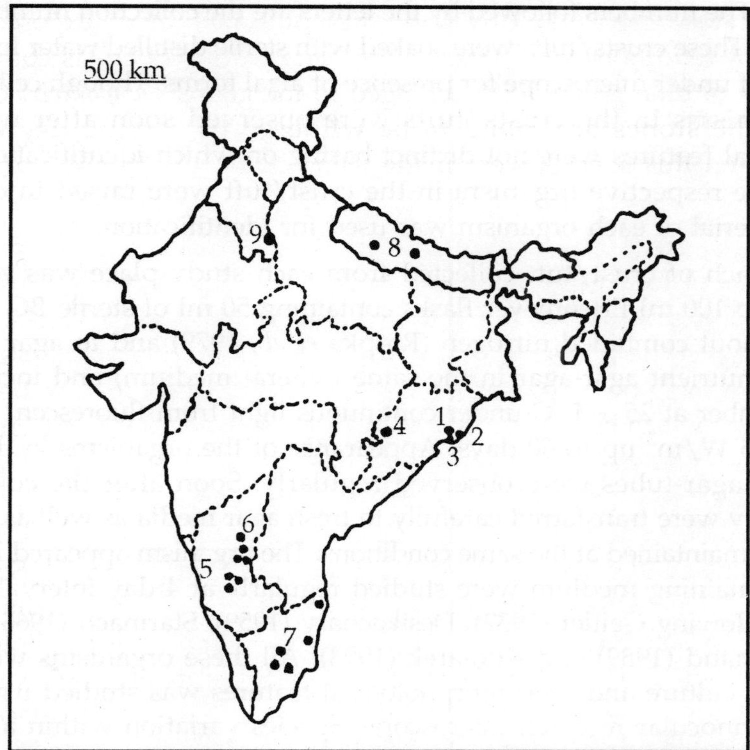

Fig. 29: Map showing location of the study sites in various regions of India and Nepal. India -A, Orissa state: 1. Bhubaneswar, 2. Konark, 3. Puri, 4. Paikamal; B, Tamilnadu state: 1. Mahabalipuram, 2. Kancheepuram, 3. Chidambaram, 4. Tiruchirappalli, 5. Tanzavur, 6. Madurai; C. Karnatak state: 1. Badami, 2. Pattadakal, 3. Aihole, 4. Sravanabelagola, 5. Belur, 6. Helebid; D. Delhi; E. Nepal - 1. Bhaktapur, 2. Patan.

73′ to 76° 18′ E); Qutub Minar, Delhi (28° 65′ N 77° 11′ E) and Krishna temple, Patan; Siddhi Laxmi and Vatasala Durga temple, Bhaktapur are located in Nepal (27° 78′ to 27° 80′ N, 85° 28′ to 85° 32′ E).

5.1.1.2 Organisms and growth conditions

The blackish brown crusts and tufts were collected from the rock surfaces of the temples and monuments of the study places during summer months (March-May, 1994 and 1995) by gently scrapping using sterile blades and needles and stored in screw cap specimen bottles. The crusts/tufts collected from different locations have been maintained at Utkal University each assigning a collection number. The first letter of the collection number denotes the state, second the city, third the site of collection and fourth the genus of blue-green alga which was the dominant component. The numbers followed by the letters are the collection number and date of collection. These crusts/tufts were soaked with sterile distilled water for few hours and observed under microscope for presence of algal forms. Though cells/filaments of the organisms in the crusts/tufts were observed soon after wetting, the morphological features were not distinct basing on which identification could be made. So the respective organism in the crust/tuft were raised in culture and unialgal material of each organism was used for identification.

A pinch of crust/tuft collected from each study place was washed and transferred to 100 ml Erlenmeyer flasks containing 50 ml of sterile BG 11 medium with or without combined nitrogen (Rippka *et al.*, 1979) and to agar plates (1%, w/v, Difco-nutrient agar-agar in the same mineral medium) and incubated in a growth chamber at $25 \pm 1° C$ under continuous light from fluorescent tubes at an intensity 12.5 W/m^2 up to 60 days. Appearance of the organisms in the enriched culture and agar-tubes were observed regularly. Soon after the colonies were obtained, they were transferred carefully to fresh agar media as well as to enriched cultures and maintained at the same conditions. The organism appeared in the liquid and agar containing medium were studied regularly at 4 day intervals and were identified following Geitler (1932), Desikachary (1959), Starmach (1966), Rippka *et al.* (1979), Anand (1989) and Komárek (1993). All these organisms were isolated into unialgal culture and their morphological features was studied using a MEIJI TH-ML-05 trinocular research microscope. Species variation within a genus was studied by their morphological features, rate of photosynthetic oxygen evolution and respiratory oxygen uptake and nitrogenase activity, and also by determination of growth rate in unialgal culture. The organisms were grown in cotton stoppered hardglass test tubes containing sterilized BG 11 medium (KNO$_3$ was omitted for heterocystous forms) in a culture room. At every alternate day growth was measured as the absorbance of the homogenized culture suspension at 760 nm in a Hitachi U-2000 UV-Vis spectrophotometer up to 30 days as per the method described earlier (Adhikary, 1983). Growth rate of each organism was determined and expressed as generation time in hours.

5.1.1.3 Measurement of photosynthesis and respiration

Photosynthetic oxygen evolution and respiratory oxygen uptake was measured in a Clark type oxygen electrode (Hansatech, UK) and a Systronics 1501 strip chart recorder. Homogenized suspension of a known quantity of cells was transferred to a reaction vessel fitted with an outer jacket for water circulation and temperature maintenance (25° C). To avoid reduction at the cathode and to maintain stable oxygen gradient across the membrane, the suspension of the organism was stirred with a magnetic stirrer. Light was provided by a Xenophot 100 W halogen lamp. The signal from the cathode was fed through a preamplifier to increase the efficiency of recording.

5.1.1.4 Measurement of nitrogenase activity (acetylene reduction assay)

For measurement of nitrogenase activity (acetylene reduction assay, ARA) of the organisms two ml of air was removed from each experimental tube containing known quantity of the blue green algal cells in BG 11 medium, capped tightly with subha seals and then same quantity of acetylene was injected followed by incubation under fluorescent light for a known period (4 h). One hundred μ l of the gasphase was removed in a Hamilton gas tight syringe and injected to the injector port of an Aimil Nucon 5765 model gas chromatograph fitted with parapak T SS column (80-100 mesh; carrier gas nitrogen, 30 ml/min; column temperature, 100° C; injector temperature 110° C, detector temperature 120° C). Quantity of acetylene reduced to ethylene was calculated using external ethylene standard (105 VPM ethylene in Argon; EDT research, 14 Trading estate Road, London) and the amount was expressed as n mol ethylene evolved per μ g chlorophyll-*a* of the cells per hour.

5.1.1.5 Mode of UV-C irradiation

For UV-C irradiation, a general electric Germicidal lamp (American Ultraviolet Co., USA; with its main emission at 254 nm and the radiation dose of about 85 ergs mm^2 S^1 at a distance of 22 cm) was used. A homogeneous suspension of the organism (absorbance at 760 nm=0.1) was pooled in a 75 mm diameter Petri dish and irradiated with UV-C for up to 48 h. During irradiation with UV in absence of white light, the suspension was constantly stirred by a magnetic stirrer. To study the survival of *T. byssoidea* and *Tolypothrix* sp. UU2434, isolated from water-logged rice fields (Nayak *et al.*, 1996) after exposure to UV-C, aliquots of 0.1 ml of irradiated samples were removed after certain exposure time (viz. 1,3,10,15,20,30 min and 1,6,,12,18, 24, 36,48 h) and subsequently transferred to 75 mm petri dishes containing agar (1% in BG 11 medium, w/v). The plates containing irradiated filaments were incubated in the dark for 24 h and then kept under fluorescent light. After 30 days of incubation under continuous light, the number of colonies of each plate were counted, and the per cent survival was computed taking the number of colonies in the control plate as 100%.

5.1.1.6 Extraction of pigments

For extraction of pigments, equal amounts of the organism from all of the experimental cultures (fresh weight=50 mg) were used. Pigments were extracted with 5 ml of 90% methanol (v/v) at room temperature for 2 h followed by incubation at 60° C for 2 min in dim light. Absorption spectra of the supernatant were recorded using a Hitachi 150-20 UV-visible spectrophotometer. For studying the presence of scytonemin and mycosporine amino acid (MAA)-like compounds in the crusts, equal amounts of the crusts (30 mg) before and after wetting were extracted with 20% (v/v) methanol at 45° C for 2 h or with 90% (v/v) methanol at 25° C for 3 h respectively. The contents were then centrifuged at 4000 rpm, 30 min and the supernatant was taken in a 1 cm quartz cuvette for measurement of absorption spectra in a Hitachi U2000 UV-Vis spectrophotometer.

5.1.1.7 Isolation of sheath

Sheath fractions of *Tolypothrix byssoidea* were isolated by subjecting the filaments to ultrasonication as per the procedure followed earlier (3.5.1.2).

5.1.1.8 In *vivo* radiolabelling, electrophoresis and autoradiography of proteins

Stress induced proteins were detected by *in vivo* radiolabelling with [^{35}S] methionine. *T. byssoidea* harvested in the mid exponential growth phase and the dried crusts soon after wetting in BG 11 medium were used for stress induction experiments. Heat treatment involved exposure to 50° C (1h). UV-C radiation stress was applied for 1 h at the fluency rate of 9 J.m^2 S^{-1} at a distance of 8 cm (15 W Phillips T-UV lamp as UV source). Ionizing radiation stress of 100 kr from Cobalt-60 source (Gamma radiation cell 220, AECL, Canada) was also applied to the crust and the organism. One ml concentrated culture suspensions and the treated crusts were labelled with 50μ Ci [^{35}S] methionine (specific activity, 1000-3000 Ci/m mole) for the last 5 min of exposure to each stress followed by rapid centrifugation and washing of the cells in BG 11 medium. The cells were then lysed by boiling for 5 min in a cracking buffer containing 125 mM Tris-Hcl (pH, 6.8), 5% (w/v) 2-mercapto-ethanol, 2% (w/v) SDS, 20% glycerol, 20 mM sodium azide, 1 mM phenyl methyl sulphonyl fluoride, 20 mM EGTA and 0.002% (w/v) bromophenol blue. The cell extracts were centrifuged at 14,000 rpm for 10 min to remove the insoluble cell debris as a pellet. Incorporation of the label into proteins was estimated by TCA precipitation of 1 μ l of the above cell lysate, on a filter disc, which was washed in ethanol, air dried and counted in a liquid scintillation spectrometer (Packard Tri-Carb-3255). For proper comparison, equal amounts of radioactivity (based on TCA-precipitable counts) were loaded on to gels.

Proteins were electrophoresed on 5 to 14% polyacrylamide SDS linear gradient slab gels (0.75 mm thickness) in a vertical system (Hoefer Scientfic Instruments, San Fransisco, California) overlaid with a stacking gel. The gradient gel was prepared by mixing 9 ml each of freshly prepared solutions A and B through a gradiant mixture. Solution A consisted of 0.375 M Tris/Hcl (pH 8.8) and the following components (on a weight-to volume basis): sucrose, 20%, acrylamide, 15%; bisacrylamide, 0.35%; SDS, 0.1%; and ammonium peroxydisulphate (APS), 0.035%. Solution B contained 0.375M Tris/Hcl (pH 8.8) and the following components (on a weight-to-volume basis): acrylamide, 5%; bisacrylamide, 0.135%; SDS, 0.1%; and APS, 0.035%. TEMED (N′, N′, N′, N′-tetramethyl ethylenediamine) was added, to both solutions A and B, at 0.54 μ l/ml just prior to gradient formation. The stacking gel contained 0.125 M Tris/Hcl (pH 6.8) and the following components (on a weight-to-volume basis): acrylamide, 3.75%; bisacrylamide, 0.135%; SDS, 0.1%; APS, 0.075%; and TEMED, 0.75 μ l/ml. Electrophoresis buffer contained 20 mM Tris-glycine (pH 8.3) and SDS (0.1%, w/v). Molecular weight calibration kits

(Sigma Chemical Co., St. Louis, MO) were used as molecular weight makers on all the gels.

Electrophoresis was carried out at constant voltage overnight (16 h) at 40 volts followed by 200 V, 1.5 h. Alternatively, day time runs at 100 V, 1 h followed by 200 V, 3 h were employed. The gel was stained with Comassie Brilliant Blue R-250 to visualise the molecular weight markers and destained. The gel was then vacuum dried at 80° C, for 1 h and exposed to X-ray film (INDU, Mumbai, India) for autoradiography. Typically 24 h exposure was given for 80,000 TCA precipitate counts per lane.

5.1.2 Results and discussion

5.1.2.1 Survey of epilithic blue green algae from temples and monuments

Many of the temples and monuments in different regions of India look blakish-brown due to development of microbial crusts and/or tufts on their rock surfaces (Fig. 30). Such crusts/tufts are observed on the exposed rock surface of temples, in the fissure of rocks, rock joints and even inside the sactum sanctorum where virtually little or no light is available (Fig. 30 a-d). Invariably three types of crusts and/or tufts were observed on the exposed rock surface of the temples: (i) Blackish coloured crust, tightly adhering to the sand stones in the shadded places. It was very difficult to remove these crusts from the rock surfaces even using a scalpel or blade (Fig. 31 a). At some locations epilithic crusts formed a coloured vegetation tightly adhering to the rocks, and on removing with pressure, loosening of rock below the crust and formation of petina was clearly marked (Fig. 31 b). These crusts invariably are composed of *Gloeocapsa* or *Gloeocapsopsis* (Fig. 32a). (ii) On the rock monuments of the temples certain blue green algae also form a thick mat and grow like a tuft which can be easily removed or peeled off (Fig. 31 c,d). The component microorganisms were also blue green algae belonging to *Phormidium* (Fig.32 b), *Lyngbya* (Fig. 32 c), *Plectonema* (Fig. 32 d) or *Tolypothrix* (Fig. 32 e,f; 33 a-d). (iii) On the exposed rock surface of Sun temple, Konark (Fig. 34 a) a particular species of *Tolypothrix* (Fig. 33 e) form blackish crust during summer months which is difficult to remove by fingers but can be obtained using fine needles (Fig. 34 b). This organism when occur as crust imparts blackish appearance to this "World Heritage Monument" (Fig. 34 c) hence probably known as "Black Pagoda". On the onset of monsoon, the crust imbibes water, swells and changes its appearance to greenish/bluish green (Fig. 34 d). At this stage the crust transformed into a greenish mat and can be easily removed by fingers. On removing all these crusts/ tufts during monsoon, often sand particles were found attached to these structures which shows that these microbial crusts/tufts can wither rock, loosen rock joints, eventually take part in toppling of the monuments.

Fig.30 - a-d: Blackish-brown crust/tuft as it occur on the exposed rock surface of Lingaraj temple complex, Bhubaneswar (a-b) and Madhav temple, Niali (near Konark) (c). Close view of the temple walls showing occurrence of crust/tuft at the rock joints which possibly is the cause of withering and detachment of rocks (d).

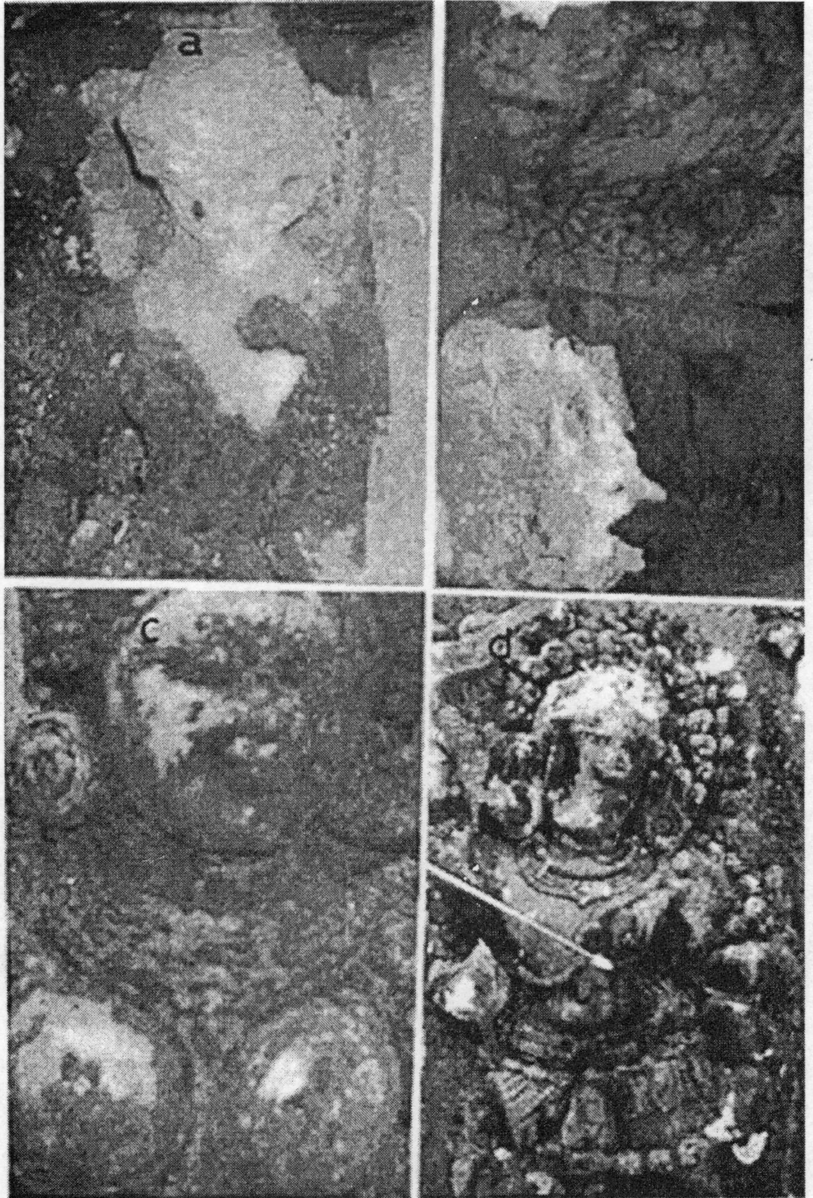

Fig. 31 a-d: Photographs showing different types of crust/tuft occurring on the rock surface of temples and monuments: a. Blackish coloured crust, tightly adhering on a sand stone monument at Madhav temple, Niali, Orissa state. b. Evidence of withering on the temples showing petina formation on the rock surfaces and penetration of blue green algae into the rocks. Blackish-brown tuft/mat on the rock surface of a monument at Yogini temple, Hirapur (near Bhubaneswar) (c), and Kedar-Gouri temple, Bhubaneswar (d)

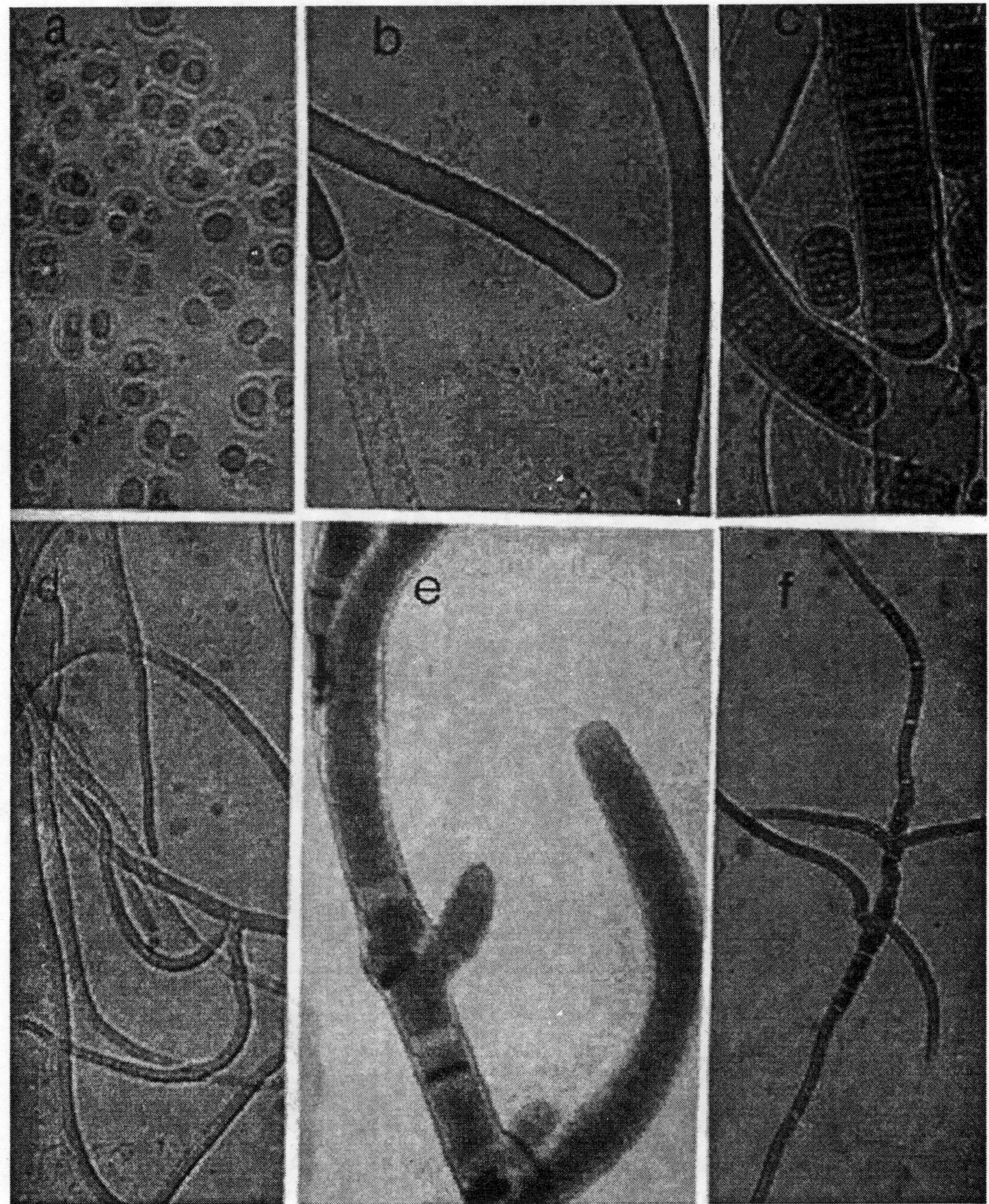

Fig. 32 a-f: Light microscopic photograph of various species of blue green algae which were the dominant component in the crust and tuft on the rock surface of various temples and monuments. a. *Gloeocapsopsis* sp. UU 515151 (x 732), b. *Phormidium* sp. UU 511160 (x732), c. *Lyngbya* sp. UU 53160 (x 732), d. *Plectonema* sp. UU 519165 (x 366), e. *Tolypothrix* sp. UU 512173 (x 732), f. *Tolypothrix* sp. UU 512176 (x 366).

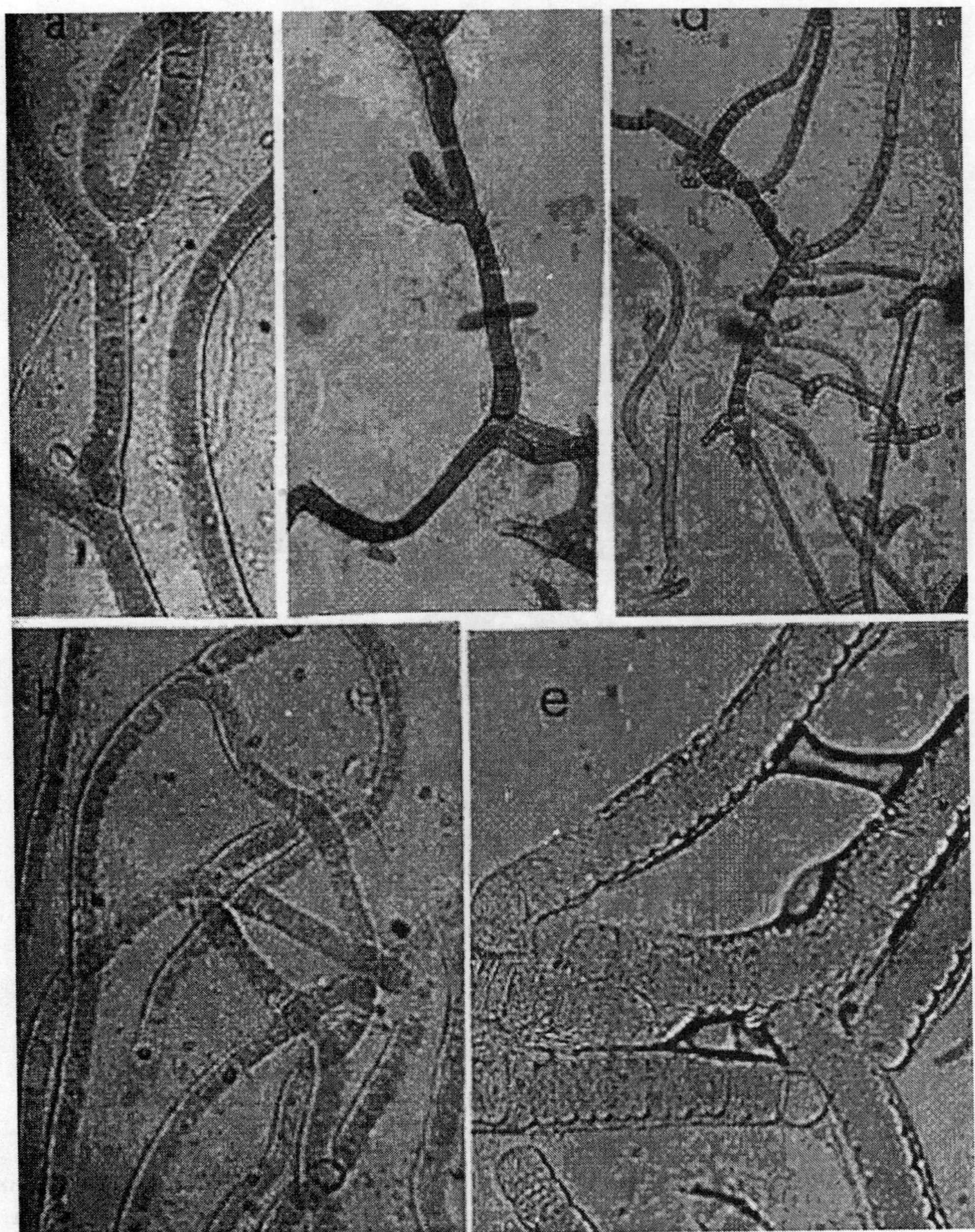

Fig. 33 a-e: Light microscopic photograph of various species of blue green algae which were the dominant component in the crust and tuft on the rock surface of various temples and monuments. a. *Tolypothrix* sp. UU 53171 (x 732), b. *Tolypothrix* sp. UU 53174 (x732), c. *Tolypothrix* sp. UU 51175 (x 366), d. *Tolypothrix* sp. UU 57172 (x 366), e. *Tolypothrix byssoidea* UU 53170 (x 732).

Fig. 34: (a) Photograph showing the Sun temple, Konark. (b) Blackish-brown crust of blue green algae on the rock surface of the temple. (c) The crust impart blackish colouration to the temple; workers of Archeological survey of India removing the crusts from the rock surfaces as a part of the routine restoration activity. (d) The blackish-brown crust becoming bluish-green tuft on the rock surface of the temple after receiving monsoon rain.Diagramatic representation of the crusts/tufts as occur on the exposed rock surface of temples during summer months and soon after receiving couple of monsoon rain is given in the illustration (Fig. 35).

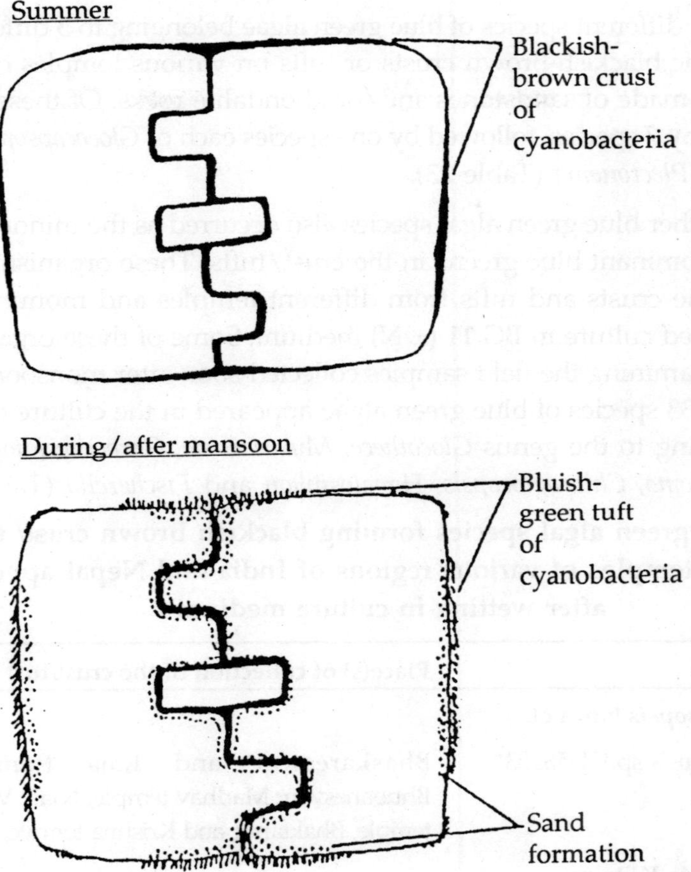

Fig. 35. Diagrammatic representation of the of occurrence of blackish-brown crust on the rock surface of the temples during summer which becomes bluish green tuft after receiving water during monsoon. The pattern of fixation of rocks on the temples, occurrence of crusts/tufts on the joints and surfaces, and sand formation below the tufts has also been shown.

Survey of epilithic blue green algae occurring on 32 different magnificent temples and monuments of India and Nepal (Tripathy *et al.*, 1997) revealed that only blue green algal forms occurred either as crust or tuft even during mid-summer months when the temperature on the rock surface goes beyond 60° C. Soon after wetting of the crusts/tufts in distilled water the organisms belonging to *Gloeocapsopsis, Lyngbya, Phormidium, Plectonema* or *Tolypothrix* appeared which were observed under microscope. Invariably *Gloeocapsopsis* or some *Tolypothrix* members occurred in the crusts where as the tufts harboured *Lyngbya, Plectonema* or *Tolypothrix* species (Fig. 33, 34).

Totally 11 different species of blue green algae belonging to 5 different genera form characteristic blackish-brown crusts or tufts on various temples of India and Nepal which are made of sandstones and/or khondalite rocks. Of these *Tolypothrix* was represented by 7 species, followed by one species each of *Gloeocapsopsis, Lyngbya, Phormidium* and *Plectonema* (Table-23).

Several other blue green algal species also occurred as the minor component along with the dominant blue greens in the crust/tufts. These organisms appeared after culturing the crusts and tufts from different temples and monuments up to 60 days in enriched culture in BG 11 ($\pm$ N) medium. Some of these organisms were also visible on examining the field samples collected soon after monsoon rain (July-August). Totally 33 species of blue green algae appeared in the culture of the crusts and tufts belonging to the genus *Gloeothece, Myxosarcina, Chroococcidiopsis, Nostoc, Calothrix, Plectonema, Chlorogloeopsis, Hapalosiphon* and *Fischerella* (Table-24).

Table 23 : Blue green algal species forming blackish brown crust/ tuft on the rock surface of temples of various regions of India and Nepal appeared soon after wetting in culture medium .

Sl No.	Organism	Place(s) of collection of the crust/tuft
	Gloeocapsopsis Nová ek	
1.	*Gloeocapsopsis* sp.UU58153*	Bhaskareswar and Raja Rani temples, Bhubaneswar; Madhav temple, Niali; Vatsala Durga temple, Bhaktapur and Krishna temple, Patan, Nepal
	Phormidium Kützing	
2.	*Phormidium* sp.UU511160	Brahmeswar and Lingaraj temples, Bhubaneswar; Virupaksha temple, Pattadakal; Meenakshi temple, Madurai; Rock Fort and Sri Rangaswamy temple, Tiruchirapalli; Bhutanath cave temple, Badami.
	Lyngbya Agardh	
3.	*Lyngbya* sp.UU53161	Parasurameswar temple, Bhubaneswar; Jagannath temple, Puri; Channekeshwara temple, Belur; Gomateshwar temple, Sravanabelagola; Natraj temple, Chidambaram; Arjun's Pennance, Mahabalipuram.
	Plectonema Thuret	
4.	*Plectonema* sp.UU519165	Khandagiri and Udaigiri ·cave temples, Bhubaneswar; Qutub Minar, Delhi; Vrihadeswar temple, Tanzavur; Rath temples, Mahabalipuram.

Sl No.	Organism	Place(s) of collection of the crust/tuft
	***Tolypothrix* Kützing**	
5.	*Tolypothrix* sp.UU53170	Sun temple, Konark; Jagannath temple, Puri; Lingaraj, Bramheswar, Megheswar, Bhaskareswar, Rameswar and Khandagiri cave temples, Bhubaneswar; Madhav temple, Niali.
6.	*Tolypothrix* sp.UU53171	Udaigiri cave temple, Bhubaneswar; Jagannath temple, Puri.
7.	*Tolypothrix* sp.UU513172	Raja Rani and Vaital temples, Bhubaneswar; Ladkhan temple, Aihole; Virupaksha temple, Pattadakal.
8.	*Tolypothrix* sp.UU512173	Mukteswara and Kedar Gouri temples, Bhubaneswar
9.	*Tolypothrix* sp.UU511174	Lingaraj temple, Bhubaneswar; Kailashnath temple, Kancheepuram; Vrihadeswar temple, Tanzavur; Nrusinghanath temple, Paikamal
10.	*Tolypothrix* sp.UU55175	Hoysaleswara temple, Helebid; Channekeshwar temple, Belur.
11.	*Tolypothrix* sp.UU512176	Kedar Gouri temple, Bhubaneswar; Jogin temple, Hirapur

Table-24: **Blue green algal species appeared in culture after cultivating crust/ tuft from rock surface of temples of India and Nepal in BG 11±N medium at 25±1° C under 12.5 W/m^2 light intensity up to 60 days.**

Sl No.	Organism	Place(s) of collection of the crust/tuft
	***Gloeocapsopsis* Nová ek**	
1.	*Gloeocapsopsis* sp. UU515151	Brahmeswar temple, Bhubaneswar;
	Gloeothece Nägeli	Bhutanath cave temple, Badami; Meenakshi temple, Madurai.
2.	*Gloeothece* sp.UU57152	Bhutanath cave temple, Badami; Meenakshi temple, Maduri.
	***Myxosarcina* Geitler**	
3.	*Myxosarcina* sp.UU514154	Megheswar and Raja Rani temples, Bhubaneswar; Virupaksha temple, Pattadakal; Ladkhan temple, Aihole
	***Chroococcidiopsis* Geitler**	
4.	*Chroococcidiopsis* sp.	Kedar Gouri and Rameswar temples,

Sl No.	Organism	Place(s) of collection of the crust/tuft
	UU512155	Bhubaneswar; Madhav temple, Niali.
5.	*Chroococcidiopsis* sp.	Rock Fort temple, Tiruchirapalli; Arjun's
	UU57156	Penance, Mahabalipuram; Bhutanath cave temple, Badami.
	***Plectonema* Thuret**	
6.	*Plectonema* sp.UU57162	Rath temples, Mahabalipuram; Khandagiri and Udaigiri cave temples, Bhubaneswar; Meenakshi temple, Madurai
7.	*Plectonema* sp.UU58163	Siddhi Laxmi and Vatsala Durga temple, Bhaktapur, Nepal.
8.	*Plectonema* sp.UU56164	Natraj temple, Chidambaram; Virupaksha temple, Pattadakal; Kailashnath temple, Kancheepuram.
	***Nostoc* Vaucher**	
9.	*Nostoc* sp.UU57157	Vrihadeswar temple, Tanzavur; Meenakshi temple, Madurai; Natraj temple, Chidambaram; Arjuna's Penance, Mahabalipuram.
10.	*Nostoc* sp.UU55158	Channekeshwar temple, Belur.
11.	*Nostoc* sp.UU52182	Sun temple, Konark.
12.	*Nostoc* sp.UU513183	Bhaskareswar, Raja Rani and Kedar Gouri temples, Bhubaneswar.
13.	*Nostoc* sp.UU52184	Sun temple, Konark.
14.	*Nostoc* sp.UU512185	Mukteswar and Khandagiri cave temples, Bhubaneswar; Vatsala Durga temple, Bhaktapur, Nepal.
15.	*Nostoc* sp.UU512186	Kedar Gouri and Parasurameswar temples, Bhubaneswar.
16.	*Nostoc* sp.UU611187	Lingaraj temple, Bhubaneswar.
17.	*Nostoc* sp.UU62188	Lingaraj temple, Bhubaneswar; Sun temple, Konark.
18.	*Nostoc* sp.UU62189	Sun temple, Konark.
19.	*Nostoc* sp.UU62190	Sun temple, Konark; Nrusinghanath temple, Paikamal.
20.	*Nostoc* sp.UU512191	Mukteswar, Kedar Gouri and Rameswar temples, Bhubaneswar.
21.	*Nostoc* sp.UU53192	Jagannath temple, Puri; Bhutanath cave temples, Badami; Meenakshi temple, Madurai.

Sl No.	Organism	Place(s) of collection of the crust/tuft
22.	*Nostoc* sp.UU516193	Bhaskareswar temple, Bhubaneswar.
23.	*Nostoc* sp.UU58194	Krishna temple, Patan, Nepal.
	***Calothrix* Agardh**	
24.	*Calothrix* sp.UU512166	Mukteswar and Megheswar temple, Bhubaneswar.
25.	*Calothrix* sp.UU519167	Khandagiri cave temples,Bhubaneswar; Hoysaleswara temple, Helebid; Bhutanath cave temples,Badami.
26.	*Calothrix* sp.UU55168	Hoysaleswar temple, Helebid.
27.	*Calothrix* sp.UU512169	Parasurameswar temple, Bhubaneswar.
	***Chlorogloeopsis* Mitra and Pandey**	
28.	*Chlorogloeopsis* sp. UU57178	Vrihadeswar temple, Tanzavur.
	***Hapalosiphon* Nägeli**	
29.	*Hapalosiphon* sp.UU512177	Kedar Gouri temple, Bhubaneswar.
30.	*Hapalosiphon* sp.UU57179	Rath temple, Mahabalipuram; Rock fort temple, Tiruchirapalli.
	***Fischerella* Gomont**	
31.	*Fischerella* sp.UU58159	Vatsala Durga temple, Bhaktapur, Nepal.
32.	*Fischerella* sp.UU512180	Parasurameswar temple, Bhubaneswar.
33.	*Fischerella* sp.UU58181	Siddhi Laxmi temple, Bhaktapur; Krishna temple, Patan,Nepal. * Isolate number was assigned after isolation and culture of each organism in unialgal state; UU, Utkal University.

All these 44 strains of blue green algae were isolated in unialgal culture. Generation time, photosynthetic oxygen evolution and dark oxygen uptake by the cells and nitrogenase activity under aerobic conditions of each organism is presented in Table-25.To analyse the species variation within a genus, each organism was cultured in a defined medium (BG 11 ± N) and unialgal cultures of each species was used for morphometric analysis. On the basis of morphological features, the epilithic cyanobacterial forms under each genus were assigned with different isolate numbers, identified up to species level and maintained at Utkal University.

SYSTEMATIC ACCOUNT

1. *Gloeocapsopsis dvorakii* (NOVACEK) KOMAREK & ANAGN. UU 58153, UU 515151.

Colonies spherical to oval, blackish-brown or yellowish green, 9.2-80.0 film broad, 6.9-20.7 µm long; sheath yellowish green colour in the outer regions and colourless inside, 4.6-9.2 flm thick; cells spherical, 4.6-9.2 µm diameter, 2-4 cells, rarely 16 cells in colonies.

Collection no.: OBBG-7, 8, 31. 07.1994; OBRG-5, 31. 07. 1994; ONMG-14, 11. 01. 1994; KBBG-24, 20. 07. 1994; TMMG-23, 01. 12. 1994.

Dominant organism in the algal crusts, collected from the rock surface of Bhaskareswar and Raja Rani temples, Bhubaneswar (Orissa); Madhav temple, Niali (Orissa). The organism appeared soon after wetting of the crust in BG 11 medium. This organism also appeared in the enriched culture of algal crusts collected from the rock surface of Brahmeswar temple, Bhubaneswar (Orissa), Bhutanath cave temple, Badami (Karnatak); Meenakshi temple, Madurai (Tamil Nadu) (Fig. 36.1).

2. *Gloeothece rupestris* (LYNGB.) BORNET, UU 57152

Colonies oval to sub-globose, greenish brown, 6.9-11.5 µm broad, 4.6-9.2 µm long; sheath thin, colourless; cells oval, 6.9-9.2 µm long, 4.6-9.2 µm broad, two cells together in oval colonies.

Collection no.: TMMG-23, 01. 12. 1994; KBBG-24, 20. 07. 1994.

Appeared in the culture of algal crusts collected from the rock surface of Meenakshi temple, Madurai (Tamil Nadu); Bhutanath cave temple, Badami (Karnatak) (Fig. 36.2).

3. *Myxosarcina spectabilis* GEITLER, UU 514154

Cells in groups of 4 or 8, sometimes single, divide in three planes, cells blackish brown, densely packed or variously pressed cells, 4.9-6.9 µm broad, 2.3-6.9 µm long, cubical cells aggregates; sheath thin, colourless.

Collection no.: OBMM-6, 31. 07. 1994;OBRM-5, 31.07.1994; KPVM-25, 20.07.1994; KALM-26, 20. 07.1994.

Appeared in the culture of the algal crusts collected from the rock surface of Megheswar and Raja Rani temples, Bhubaneswar (Orissa), Virupaksha temple, Pattadakal and Ladkhan temple of Aihole (Karnatak) (Fig. 36.3).

4. *Chroococcidiopsis indica* S. C. DIXIT, UU 512155, UU 57156

Cells in groups, sometimes single, spherical, sometimes pyriform, the difference between minimum and maximum cell size was more than double, cells 4.2-9.2 µm broad, 4.2-10.6 µm long, bluish green in colour.

Collection no.: OBKC-2, 25. 11. 1994; OBRC-9, 31. 07. 1994; ONMC-14, 11. 01. 1994; TTRC-21, 29.11. 1993; TMAC-16, 27. 11. 1993; KBBC-24. 20. 07. 1994.

Appeared in the culture of algal crusts collected from the rock surface of Kedar Gouri and Rameswar temple, Bhubaneswar; Madhav temple, Niali (Orissa); Rock Fort temple of Tiruchirappalli and Arjun's Penance of Mahabalipuram (Tamil Nadu); Bhutanath cave temple, Badami (Karnatak) (Fig. 36.4).

5. *Lyngbya corticcola* BRUHL & BISWAS, UU 53161

Thallus brownish, filaments straight, loosely intertwined; cells broader than long, 13.8-23.0 μm broad, 2.3-3.5 μm long; trichomes not constricted at the cross walls; sheath lamellated, yellowish brown, 2.3 - 4.6 μm thick; apices round.

Collection no.: OPJL-12, 22. 08. 1994; OBPL-3, 25.11. 1994; KBCL-28, 18.07. 1994; KSGL-27, 18.07.1994; TCNL-19, 27. 11. 1993; TMAL-16, 27. 11. 1993.

Dominant organism in the algal crusts collected from the rock surface of Jagannath temple, Puri; Parasurameswar temple, Bhubaneswar (Orissa); Channekeshwara temple, Belur; Gomateshwar temple, Sravanabelagola (Karnatak); Nataraja temple, Chidambaram and Arjun's Pennance of Mahabalipuram (Tamil Nadu). The organism appeared soon after wetting the crust in BG 11 medium (Figs. 36.5).

6. *Phormidium trunicicola* S. L. GHOSE, UU 511160

Thallus brownish, filaments straight, loosely intertwined; cells slightly broader than long, 6.9-14.5 μm broad, 4.6-11.5 μm long; attenuated at the apices; trichomes slightly constricted at the cross walls, sheath thin, brownish, 1.0- 1.2 μm thick.Collection no.: OBLL-I, 25. 11. 1994; OBBL-7, 31. 07.1994; TMML-23, 01. 12. 1994; TTRL-21, 29. 11. 1993; TTSL-20, 29. 11. 1993; KPYL-25, 20.07. 1994; KBBL-24, 20. 07.1994.

Dominant organism in the algal crusts collected from the rock surface of Lingaraj and Brahmeswar temples of Bhubaneswar (Orissa); Meenakshi temple, Madurai, Rock Fort and Sri Rangaswamy temple, Tiruchirapalli (Tamil Nadu); Virupaksha temple of Pattadakal and Bhutanath cave temple of Badami (Karnatak). The organism appeared soon after wetting of the crust in BG II medium (Figs. 6, 33).

7. *Plectonema gracillimum* (ZOPF.) HANSG., UU 57162

Thallus leathery, bluish green, filaments densely entangled, false branched; cells small, isodiametrical to shorter than wide, 2.3-4.6 μm broad, 2.3 - 2.6 μm long; sheath colourless, 1.0- 1.2 μm thick; apices round. Collection no.: OBKP-IO, 23.08. 1994; TMMP-23, 01. 12. 1994; TMPR-17, 27. 11. 1993.

Appeared in the culture of algal crusts collected from the rock surface of Khandagiri and Udaigiri cave temples, Bhubaneswar (Orissa); Meenakshi temple, Madurai and Rath temple of Mahabalipuram (Tamil Nadu) (Fig.36. 7).

8. *Plectonema hansgirgi* SCHMIDLE , UU 56164

Thallus greenish, filaments straight, sometimes curved, false branched; cells broader than long, 6.9-8.4 µm broad, 2.3-2.9 µm long; sheath colourless, 1.0 - 1.2 µm thick; apices round. Collection no.: OKSP-II, 23. 08. 1994;KPVP-25, 20.07. 1994; TCNP-19, 27. 11. 1993; TKKP-18, 27. 11. 1993. Appeared in the culture of algal crusts collected from the rock surface of Sun temple, Konark (Orissa); Virupaksha temple, Pattadakal (Kamatak); Nataraja temple, Chidambaram and Kailashnath temple of Kancheepuram (Tamil Nadu) (Fig. 36.8).

9. *Plectonema puteale* (L. A. KIRCHNER) HANSG., UU 519165

Thallus brownish, filaments long, densely entangled, false branched frequent, branches short, slightly narrower than the main filament; cells broader than long, 2.3-6.9 µm broad, 2.3-2.9 µm long; trichomes not constricted at the cross wall; sheath colourless, 1.0-1.3 µm thick; apices round. Collection no.: OBKP-1O,23. 08. 1994;DDKP-30, 12.02. 1995; TTVP-22, 30.11. 1993; TMRP-I7, 27. 11. 1993. Dominant organism in the algal crusts collected from the rock surface of Khandagiri and Udaigiri cave temples, Bhubaneswar (Orissa); Qutub Minar (Delhi); Brihadeswara temple, Tanjavur and Rath temple of Mahabalipuram (Tamil Nadu) (Fig. 36. 9).

10. *Nostoc carneum* C. AGARDH ex BORNET & FLAHAULT, UU 56158

Thallus olive green, mucilaginous, irregularly expanded; filaments straight; cells barrel shaped, longer than broad, 2.3-4.6 µm broad, 4.6-6.9 µm long; heterocyst single, intercalary, elongate, 2.3-4.6 µm, 4.3-6.9 µm long; akintes yellowish brown, oblong, many in chains, 4.6-11.5 µm broad. 5.8-2.6 mm long.

Collection no.: KBCN-28, 18.07. 1994.

Appeared in the culture of algal crusts collected from the rock surface of Channekeshwar temple, Belur (Krnatak) (Fig. 36.10).

11. *Nostoc ellipsosporum* (DESM.) RABENH.ex BORNET & FLAHAULT, UU 52184, UU 611187, UU 62188

Thallus bluish green or greyish, gelatinous, irregularly expanded; filaments straight, long, loosely entangled; cells spherical to cylindrical, longer than broad, 3.3-5.9 µm broad, 4.6-8.9 µm long; heterocysts single, sometimes many in chains, terminal as well as intercalary, sub-spherical to ellipsoidal, 4.6

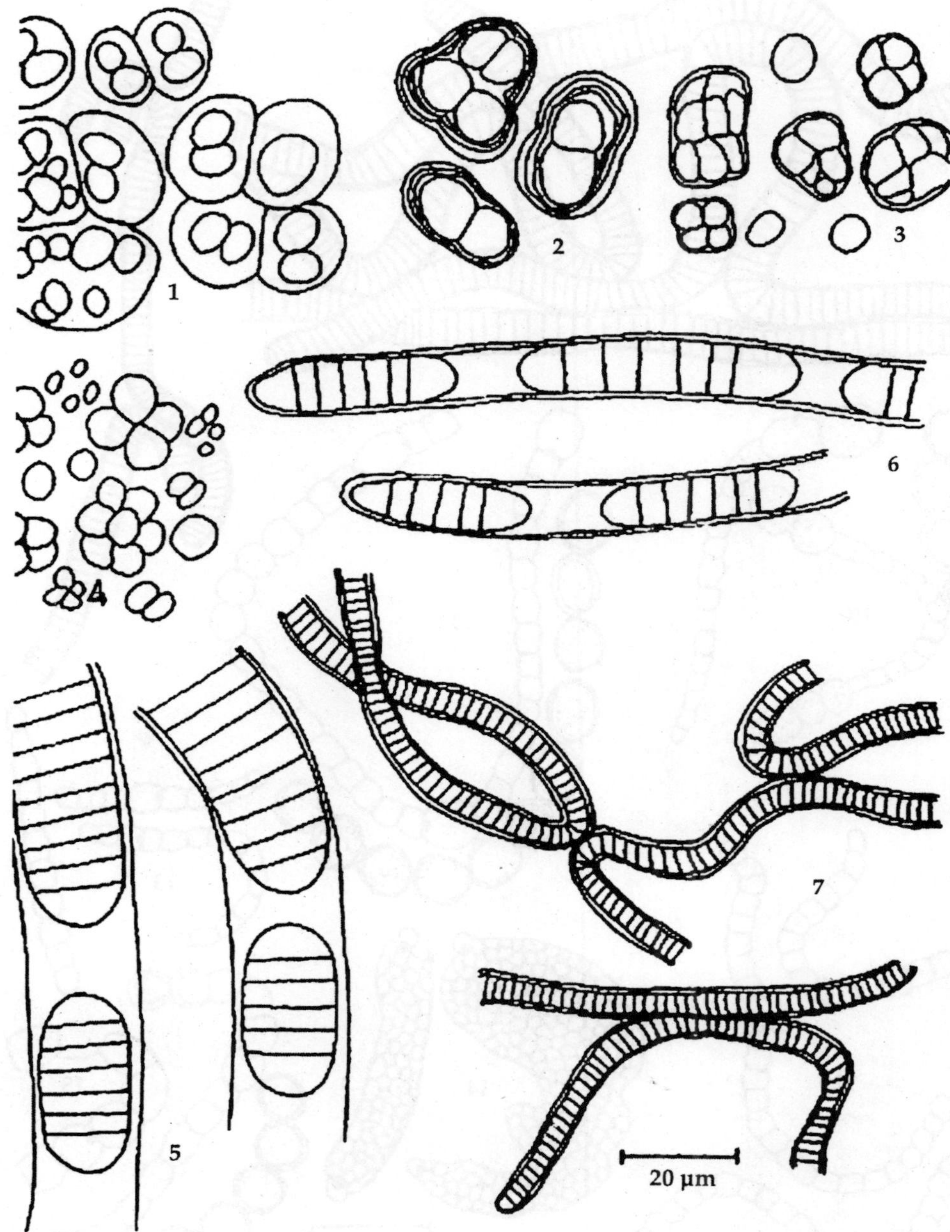

Figs.36: 1 – 7 :Various species of blue-green algae

1 - *Gloeocapsopsis dvorakii*; 2 - *Gloeothece rupestris*; 3 *Myxosarcina spectabilis*;
4 - *Chroococcidiopsis indica*; 5 *Lyngbya corticcola*; 6 - *Phormidium truncicola*;
7 *Plectonema gracillimum*

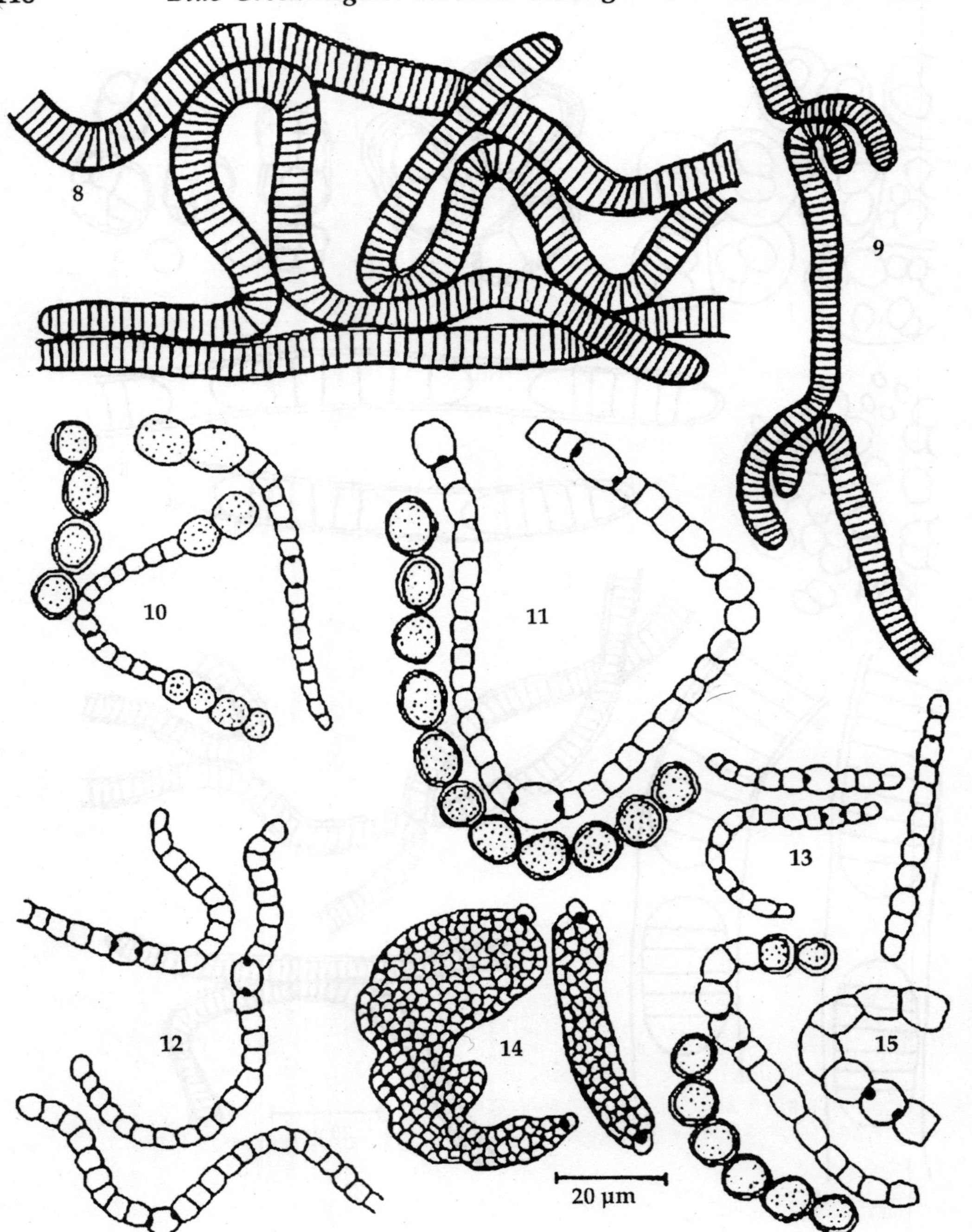

Figs. 8-15 : Various species of blue-green algae

8 – *Plectonema hansgirgi*; 9 - *P. puteale*; 10 - *Nostoc carneum*; 11 - *N. ellipsosporum*;
12 - *N.muscorum*; 13 *N. paudosum*; 14 – *N. punctiforme*; 15 - *N.spongiaeforme*

to 9.2 µm broad,4.6-13.8 µm long; akinetes oval, sometimes cylindrical, many in chains, 4.6-13.8 µm long, 5.5-10.2 µm broad, some in germinating stages.

Collection no.: OKSN-II, 31. 07. 1994; OBLN-I, 31. 07.1994.

Appeared in the culture of algal crusts collected from the rock surface of Sun temple, Konark and Lingaraj temple, Bhubaneswar (Orissa) (Fig. 36.11).

12. *Nostoc muscorum* C. AGARDH ex BORNET & FLAHAULT, UU516193

Thallus olive green, mucilaginous; filaments long; cells nearly spherical, as long as broad, 4.6-6.9 µm diameter, heterocyst single, intercalary, spherical, 4.6-7.8 ~m in diameter; akinetes not recorded.

Collection no.: OBBN-8, 31. 07.1994.

Appeared in the culture of algal crusts collected from the rock surface of Bhaskareswar temple, Bhubaneswar (Orissa) (Fig. 36.12).

13. *Nostoc paludosum* KUTZ.ex BORNET & FLAHAULT, UU 57157

Thallus blackish brown; filaments short; cells barrel shaped, longer than broad, 2.3 - 2.8 µm broad, 2.3-4.6 µm long; heterocyst single, intercalary, barrel shaped, longer than broad, 2.3-3.5 µm broad, 4.6-6.9 µm long; akinete single, sometimes many in chain, spherical, 2.3-9.2 µm in diameter.

Collection no.: TTVN-22, 30. 11. 1993; TMMN-23, 01. 12.1994; TCNN-19, 27. 11. 1993; TMAN-16, 27. 11. 1993.

Appeared in the culture of algal crusts collected from the rock surface of Brihadeswara temple, Tanjavur; Meenakshi temple, Madurai and Nataraja temple of Chidambaram (Tamil Nadu); Arjuna's Penance, Mahabalipuram (Tamil Nadu) (Fig. 36. 1 3).

14. *Nostoc punctiforme* C. AGARDH ex BORNET & FLAHAULT, UU 513183

Thallus bluish green, small dot like, united to form a mass; filaments coiled, densely entangled; cells short barrel shaped or ellipsoidal, 2.3-6.9 µm broad, 2.3-4.6 µm long; heterocyst single, terminal, subspherical to spherical, 4.6-6.9 µm broad, 2.3-4.6 µm long; akinetes spherical to oblong, 4.6-6.9 µm in diameter; sheath thin, distinct at the periphery.

Collection no.: OBRN-5, 31. 07. 1994;

OBBN-8, 31. 07.1994; OBKN-2, 31. 07.1994.

Appeared in the culture of algal crusts collected from the rock surface of Raja Rani, Bhaskareswar and Kedar Gouri temples, Bhubaneswar (Orissa) (Fig. 14).

15. *Nostoc spongiaeforme* C. AGARDH ex BORNET & FLAHAULT, UU 62189

Thallus brownish, leathery, irregularly expanded; filaments loosely entangled; cells barrel shaped to cylindrical, longer than broad, 2.3-4.6 µm broad,

4.6-9.2 µm long, heterocyst nearly spherical, sometimes longer than broad, 4.6-6.9 µm broad, 6.9-9.2 µm long; akinetes spherical, many in series, 4.6-11.5 µm in diameter.

Collection no.: OKSN-II, 31. 07. 1994.

Appeared in the culture of algal crusts collected from the rock surface of Sun temple, Konark (Orissa) (Fig. 36.15).

16. *Calothrix brevissima* G. S. WEST var. *moniliforme* S. L. GHOSE orth. mut. GEITLER, UU 55168

Thallus blackish brown; filaments short, straight, slightly attenuated; cells at the base broader than long, gradually globose, 6.9-9.2 µm broad, 4.6-6.9 µm long; trichome moniliform; heterocyst single, basal, subspherical to spherical, 4.6-6.9 µm broad, 2.3-4.6 µm long; sheath colourless, 1.0-1.2 µm thick; apices round.

Collection no.: KHHC-29, 18.07. 1994.

Appeared in the culture of algal crusts collected from the rock surface of Hoysaleswar temple, Helebid (Karnatak) (Fig. 36.16).

17. *Calothrix ghosei* BHARADWAJA, UU 512169

Thallus bluish green; filaments straight, sometimes little bent, slightly swollen at the base, gradually tapers; cells broader than long, 4.6-13.8 µm broad, 4.6-9.2 µm long; in old cultures cells densely pigmented; heterocyst single, basal, subspherical to spherical, 4.6-1 1.5 µm broad, 4.6-9.2 µm long; akinete single, developed in old cultures adjoining to the basal heterocyst, spherical, 9.2-18.4 µm in diameter; sheath colourless, 1.2-2.3 µm thick.

Collection no.: OBPC-333, 25. II. 1994.

Appeared in the culture of algal crusts collected from the rock surface of Parasurameswar temple, Bhubaneswar (Orissa) (Fig. 36. 17).

18. *Calothrix marchica* LEMMERM. var. *crassa* C. B. RAO, UU 519167

Thallus brownish; filaments long, straight, sometimes bent, slightly swollen at the base, gradually tapers; cells at the base much broader than long, 4.6-3.8 µm broad, 2.3-6.9 µm long; heterocyst single, sometimes double, basal, subspherical, 6.9-10.5 µm broad, 6.9-16.1 µm long; sheath yellowish brown, 1.2-2.3 µm thick; apices round.

Collection no.: OBCK-IO, 23.08.1994; KHHC-29, 18.07.1994; KBBC-24, 20. 07. 1994.

Appeared in the culture of algal crusts collected from the rock surface of Khandagiri cave temples, Bhubaneswar (Orissa); Hoysaleswara temple, Helebid and Bhutanath cave temple of Badami (Karnatak) (Fig. 36. 18).

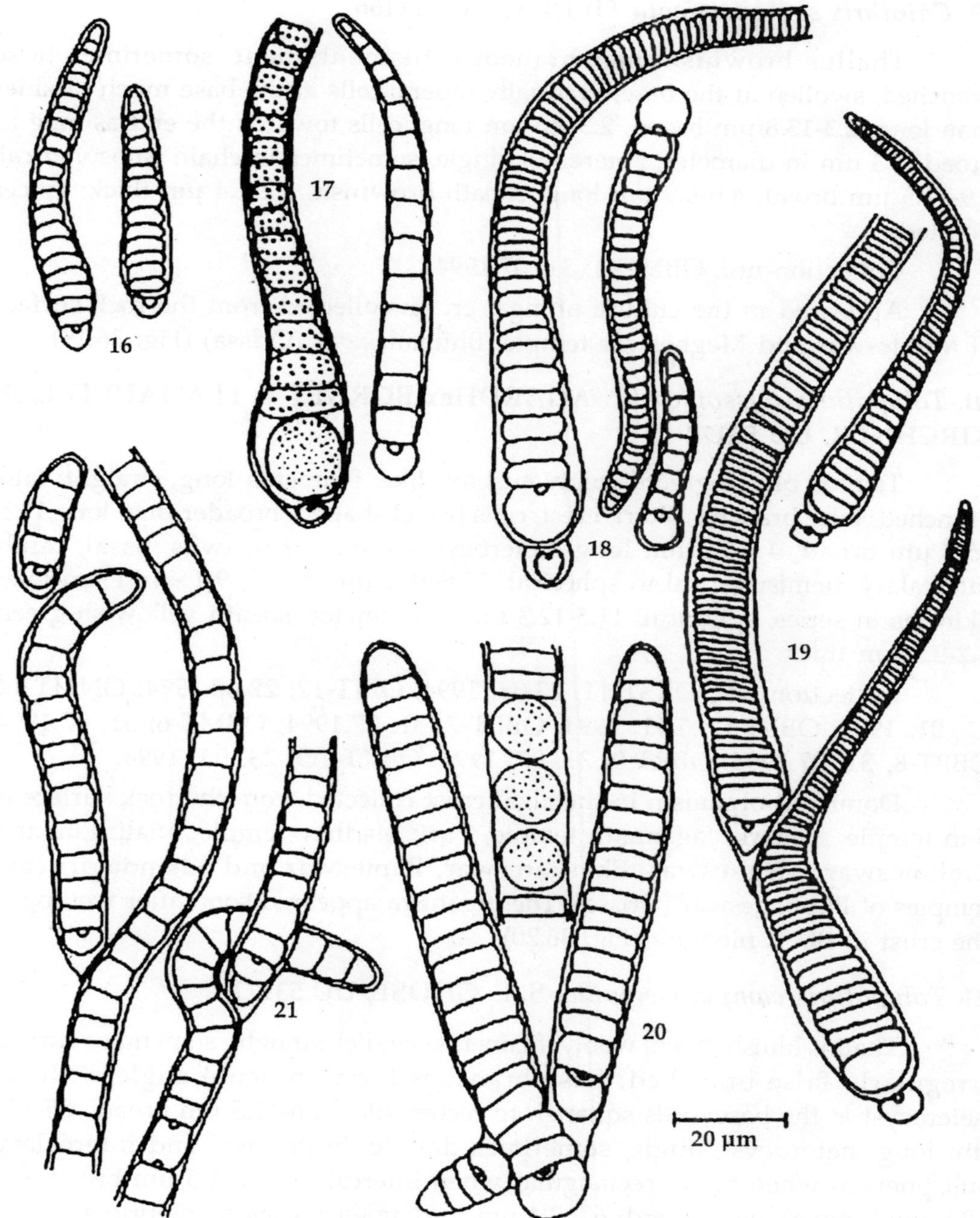

Figs. 36: 16-21 : Various species of blue green algae
16- *Calothrix brevissima*, 17- *C. ghosei*, 18- *C. marchica*, 19- *C. scytonemicola*,
20- *Tolypothrix byssoidea*, 21- *T. campylonemoides*

19. *Calothrix scytonemicola* TILDEN, UU 512166

Thallus brownish, membraneous; filaments bent, sometimes false branched, swollen at the base, gradually tapers; cells at the base much broader than long, 2.3-13.8 μm broad, 2.3-4.6 μm long, cells towards the end as long as broad, 2.3 μm in diameter; heterocyst single, sometimes in chain, subspherical, 6.9-11.5 μm broad, 4.6-6.9 μm long; sheath brownish, 1.2-1.4 μm thick; apices pointed.

Collection no.: OBMC-4, 31. 07.1994.

Appeared in the culture of algal crusts collected from the rock surface of Mukteswar and Megheswar temple, Bhubaneswar (Orissa) (Fig. 36.19).

20. *Tolypothrix byssoidea* (C. AGARDHex BORNET & FLAHAULT) L. A. KIRCHNER, UU 53170

Thallus bluish green, wooly cushion like; filaments long, straight, false branched; false branches short, erect; cells barrel shaped, broader than long,11.5-25.3 μm broad, 4.6-9.2 μm long; heterocysts single or in twos, basal, rarely intercalary, hemispherical to spherical, 13.8-16.1 μm broad, 9.2 - 16.1 μm long; akinetes in series, spherical, 11.5-12.3 μm in diameter; sheath yellowish green, 1.2-2.3 μm thick.

Collection no.; OKST-11, 23.08. 1994; OPJT-12, 22.08.1994; ONMT-14, 11. 01. 1996; OBLT-I, 25.11. 1994; OBBT-7, 31. 07.1994; OBMT-6, 31. 07.1994; OBBT-8, 31. 07.1994; OBRT-9, 31. 07. 1994; OBKT-1O, 23. 08. 1994.

Dominant organism in the algal crust collected from the rock surface of Sun temple, Konark; Jagannath temple, Puri; Madhav temple, Niali; Lingaraj, Brahmeswar, Megheswar, Bhaskareswar, Rameswar and Khandagiri cave temples of Bhubaneswar (Orissa). The organism appeared soon after wetting of the crust in BG II medium (Fig. 36.20).

21. *Tolypothrix campylonemoides* S. L. GHOSE, UU 53171

Thallus bluish green, wooly; filaments parallel, straight, sometimes curved, irregularly false branched; false branches form an acute angle with the heterocyst at the base; cells squarish to rectangular, 4.6-11.5 μm broad, 4.5-13.8 μm long; heterocyst single, sometimes double, both basal and intercalary, subspherical when basal, rectangular when intercalary, 9.2-1.5 μm broad, 9.2-13.8 μm long; akinete round, 6.9-7.4 μm in diameter; sheath colourless, 1.2-2.3 μm thick; apices round.

Collection no.: OPJT-12, 22.08. 1994; OBKT-IO, 23. 08.1994.

Dominant organism in the algal crusts collected from the rock surface of Jagannath temple, Puri, and Udaigiri cave temple, Bhubaneswar (Orissa). The organism appeared soon after wetting of the crust in BG 11 medium (Fig. 36. 21).

22. *Tolypothrix crassa* W. WEST & G. S. WEST, UU 511174

Thallus bluish green, wooly; filaments long, false branched; false branches erect; cells broader than long, 16.1-25.3 µm broad, 4.6-6.9 µm long; heterocyst single, sometimes 2 - 3 in chain of 2 or 3 together, both basal and intercalary, subspherical to cylindrical, 11.5-13.8 µm broad, 4.6-9.2 mm long; separation disc present; sheath brownish,

1.2-2.3 mm thick; apices round.Collection no.: OBLT-l, 31. 07.1994; OPNT-13, 05.12.1994; TKKT-18, 27.11. 1993; TTVT-22,30.

23. *Tolypothrix distorta* KUTZ. ex BORNET & FLAHAULT *var.penicillata* (C. AGARDH) LEMMERM., UU 512176

Thallus bluish green, wooly; filaments straight, false branched; cells broader than long, 9.2-16.1!Jm broad, 3.6-4.9!Jm long; heterocyst single, basal as well as intercalary, eIliptical to subspherical, 9.2-16.1 µm broad, 4.6-9.2 µm long; sheath brownish, irregularly broadened, 2.3-4.6 µm thick; apices round.

Collection no.: OBKT-2, 31. 07.1994; OHJT-15, 11. 01. 1996.

Dominant organism in the algal crusts collected from the rock surface of Kedar Gouri temple, Bhubaneswar and Jogin temple, Hirapur (Orissa). The organism appeared soon after wetting of the crust in BG 11 medium (Figs. 36.23).

24. *Tolypothrix nodosa* BHARADWAJA, UU 55175

Thallus brownish; filaments straight, false branched; false branches short, curved; cells barrel shaped, longer than broad, 4.6-6.9 µm broad, 6.9-9.2 µm long; heterocyst single, both basal and intercalary, subspherical to rectangular, 4.6-6.9 µm broad, 2.3-6.9 µm long; sheath colourless, 1.1-1.6 µm thick; apices round.

Collection no.: KHHT-29, 18.07.1994; KBCT-28, 18.07.1994.

Dominant organism in the algal crust collected from the rock surface of Hoysaleswara temple, Helebid and Channekeshwar temple of Belur (Krnatak). The organism appeared soon after wetting of the crust in BG II medium (Fig. 36.24).

25. *Tolypothrix rechingeri* (WILLE) GEITLER, UU 513172

Thallus brownish, wooly; filaments straight, sometimes curved, false branched; false branches short, many; cells barrel shaped, broader than long, 9.2-16.I µm broad, 6.9-9.2 µm long; trichomes constricted at the cross walls;

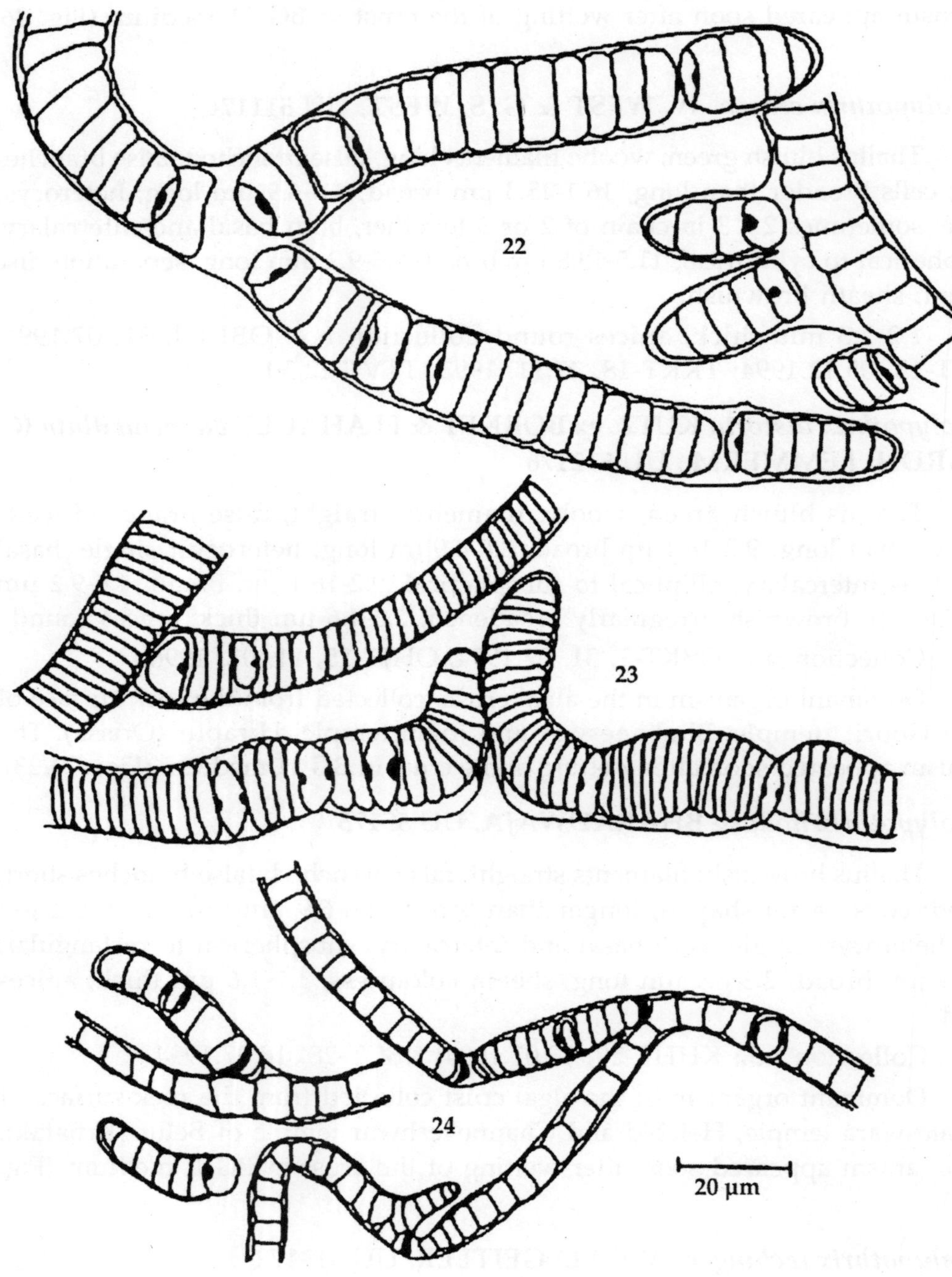

Figs: 36. 22-24 : Various species of blue-green algae
22 - *Tolypothrix crassa;* 23 - *T. distorta;* 24 - *T. nodosa*

heterocyst both basal and intercalary, subspherical to barrel shaped, subspherical when basal and barrel shaped when intercalary, 11.5-13.8 µm broad, 4.6-11.5 µm long; sheath yellowish brown, 2.3-4.6 µm thick. Collection no.: OBRT-5, 31. 07. 1994; KALT-26, 20.07. 1994; KPVT-25, 20. 07.1994.Dominant organism in the algal crusts collected from the rock surface of Raja Rani and Vaital temples, Bhubaneswar (Orissa); Ladkhan temple, Aihole and Virupaksha temple of Pattadakal (Karnatak). The organism appeared soon after wetting of the crust in BG11 medium (Fig. 36. 25).

26. *Tolypothrix scytonemoides* (GARDNER) GEITLER, UU512173

Thallus greenish brown, wooly; filaments straight, sometimes curved, false branched; false branches short, single; cells broader than long, 11.5 - 23 µm broad, 2.3-9.2 µm long; trichome constricted at the cross walls; heterocysts basal and intercalary; subspherical to ellipsoidal, basal one subspherical, intercalary one ellipsoidal, 13.8 -18.4 µm broad, 6.9-11.5 11m long; akinetes 11.5-13.8 µm in diameter; sheath yellowish brown, 2.3-9.2 µm thick; apices round.

Collection no.: OBMT-4, 31. 07.1994; OBKT-2, 25. 11. 1994.

Dominant organism in the algal crust collected from the rock surface of Mukteswar and Kedar Gouri temples, Bhubaneswar (Orissa). The organism appeared soon after wetting of the crust in BG II medium (Fig. 36.26).

27. *Chlorogloeopsis fritschii* T. MITRA & D. C. PANDEY, UU 57178

Thallus bluish green; cells showed division in two planes; filaments short, straight, 3- to 5-celled; cells cylindrical to subspherical, broader than long, 2.3-6.9 µm broad, 2.3-4.6 µm long; heterocyst single, terminal or intercalary, subspherical to spherical, 2.3-4.6 µm broad, 2.3-4.6 µm long; akinetes rounded, 4.6-6.9 µm in diameter; sheath thin.

Collection no.: TTBC-22, 31. 07.1994.

Appeared in the culture of algal crusts collected from the rock surface of Brihadeswara temple, Tanjavur (Tamil Nadu) (Fig. 36.27).

28. *Fischerella muscicola* (THUR.) GOMONT, UU 512180

Thallus yellowish green; filaments branched, two types, main filament with short barrel shaped cells; cells of lateral filament longer than broad, 2.3-6.9 µm broad, 3.0-20.7 µm long; heterocyst single, intercalary, cylindrical, longer than broad, 2.3-6.9 µm broad, 6.9-20.7 µm long; akinetes in chain, subspherical to spherical, 4.6-13.8 µm in diameter; sheath thin.

Collection no.: OBPF-3, 25. 11. 1994. Appeared in the culture of algal crusts collected from the rock surface of Parasurameswar temple, Bhubaneswar (Orissa) (Fig. 28).

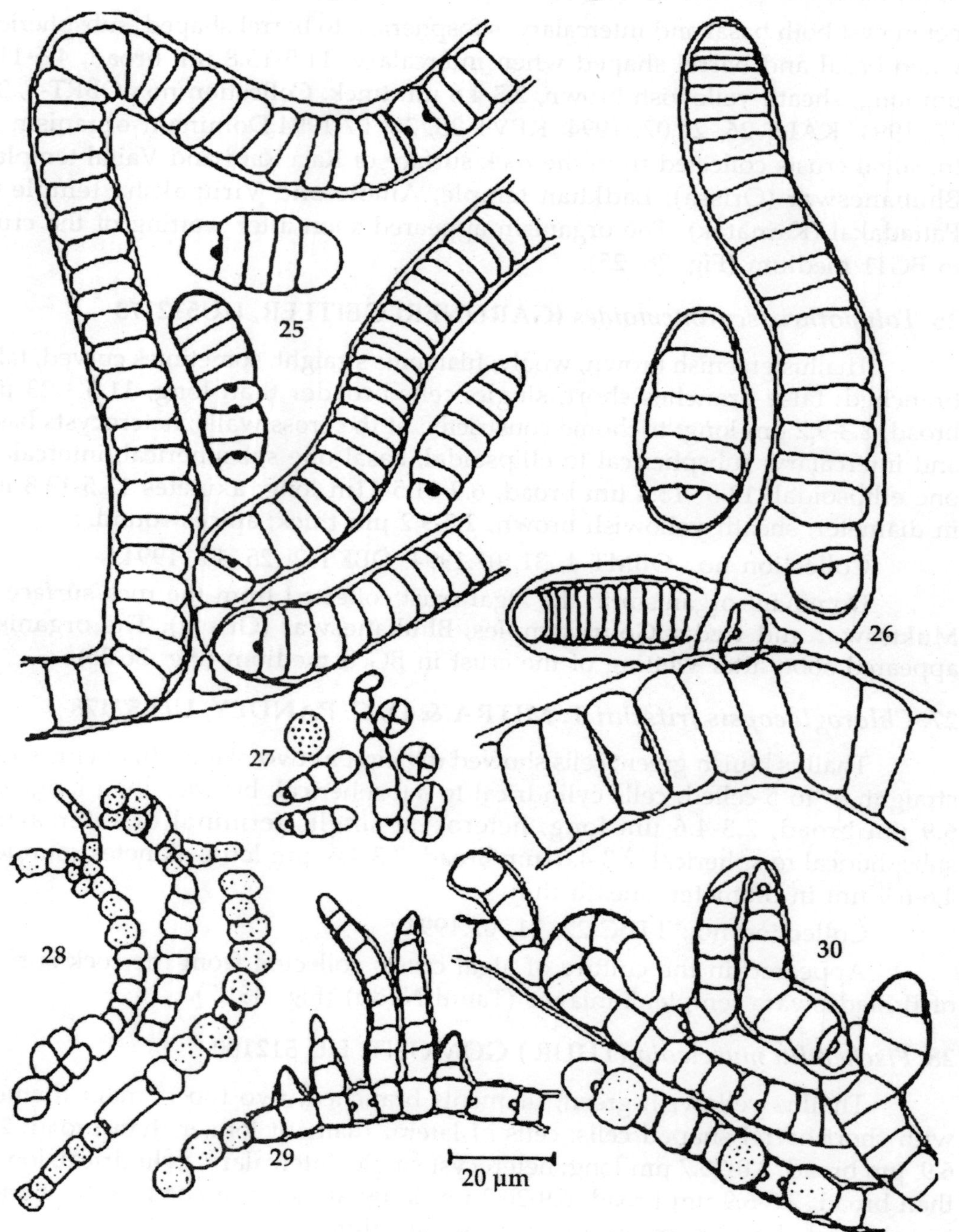

Figs: 36.25-30 : Various species of blue-green algae
25 - *Tolypothrix rechingeri;* 26 - *T. scytonemoides;* 27 *Chlorogloeopsis fritschii;*
28 - *Fischerella muscicola;* 29 *Hapalosiphon flagelliformis;* 30 - *H. stuhlmannii*

29. *Hapalosiphon flagelliformis* (SCHMIDLE) FORTI, UU 57179

Thallus bluish green, spreading from a common point; filaments densely interwoven, branched; lateral branches short, attenuated, as broad as main filament or little narrower; cells of prostrate filament barrel shaped, longer than broad, 4.6-6.9 µm broad, 7.4-9.2 l1m long; heterocyst single, intercalary, subquadrat to cylindrical, 4.6-6.9 µm broad, 7.4-9.2 µm long; akinetes many in chains, spherical, 4.6-6.9 µm in diameter; sheath colourless, thin.Collection no.: TMRH-I7, 27. 11. 1993; TTRH-21, 30. 11. 1993.

Appeared in the culture of algal crusts collected from the rock surface of Rath temple, Mahabalipuram and Rock Fort temple, Tiruchirappalli (Tamil Nadu) (Fig. 36.29).

30. *Hapalosiphon stuhlmannii* HIERON., UU 512177

Thallus bluish green, leathery; filaments long, straight, continuously branched; branches irregular, lateral, often arising on one side of the filaments, sometime false branches seen just like those of Scytonemataceae, true branches slightly attenuated; cells in single row, quadrate to ellipsoidal, longer than broad, 4.6-9.2 µm broad, 6.9-11.5 µm long; heterocysts frequent, single, sometimes in chains of 2-5, terminal as well as intercalary, spherical to rectangular, 4.6-6.9 µm broad, 6.9-9.2 µm long; akinetes yellowish brown, subspherical to spherical to ellipsoidal, larger than vegetative cells.

Collection no.: OBKH-2, 31.07.1994.Appeared in the culture of algal crusts collected from the rock surface of Kedar Gouri temple, Bhubaneswar, Orissa (Fig. 36:30).

Table-25: Growth (generation time in hours), photosynthetic oxygen evolution and respiratory oxygen uptake (µ mol O_2.mg Chl-a^{-1}.h^{-1}) and nitrogenase activity (n mol C_2H_4.µ g Chl-a^{-1}.h^{-1}.) of blue green algal species isolated from the rock surface of various temples of India and Nepal.

Sl no.	Organism	Isolate number	Generation time (h)	Photosynthetic O_2 evolution	Respiratory O_2 uptake	ARA activity (n mol ethylene.
				(µ mol O_2.mg Chl-a^{-1}.h^{-1})		(µ g Chl-a^{-1}.h^{-1}.)
1	2	3	4	5	6	7
1.	*Gloeocapsopsis* sp.	UU 58153	121	180.2	156.7	0
2.	*Phormidium* sp.	UU 511160	215	149.5	188.5	0
3.	*Lyngbya* sp.	UU 53161	230	71.9	85.9	0
4.	*Plectonema* sp.	UU 519165	164	72.1	901.6	0.25
5.	*Tolypothrix* sp.	UU 53170	210	72.9	104.3	1.11

Contd. ....

1	2	3	4	5	6	7
6.	*Tolypothrix* sp.	UU 53171	158	50.8	44.2	2.92
7.	*Tolypothrix* sp.	UU 513172	189	26.9	17.9	1.44
8.	*Tolypothrix* sp.	UU 512173	202	30.7	61.0	1.41
9.	*Tolypothrix* sp.	UU 511174	202	24.6	55.4	0.80
10.	*Tolypothrix* sp.	UU 55175	134	376.6	143.7	0.59
11.	*Tolypothrix* sp.	UU 512176	188	59.23	52.3	2.73
12.	*Gloeocapsopsis* sp.	UU 515151	290	535.1	356.3	0.44
13.	*Gloeothece* sp.	UU 57152	276	350.0	194.5	0
14.	*Myxosarcina* sp.	UU 514154	181	690.7	460.5	0
15.	*Chroococcidiopsis* sp.	UU 512155	193	285.6	158.0	0.07
16.	*Chroococcidiopsis* sp.	UU 57156	139	223.0	117.6	0.09
17.	*Plectonema* sp.	UU 55162	202	443.5	272.6	0
18.	*Plectonema* sp.	UU 58163	174	225.5	451.0	0
19.	*Plectonema* sp.	UU 52164	189	593.3	485.9	0
20.	*Nostoc* sp.	UU 57157	157	383.7	135.4	0.1
21.	*Nostoc* sp.	UU 55158	215	1013.7	337.9	0.14
22.	*Nostoc* sp.	UU 52182	136	689.7	318.3	1.55
23.	*Nostoc* sp.	UU 513183	199	901.6	541.0	4.55
24.	*Nostoc* sp.	UU 52184	133	609.9	217.8	0.73
25.	*Nostoc* sp.	UU 512185	211	739.3	324.6	0.81
26.	*Nostoc* sp.	UU 512186	103	161.1	80.6	2.93
27.	*Nostoc* sp.	UU 611187	166	530.1	132.6	4.83
28.	*Nostoc* sp.	UU 62188	151	473.6	75.2	2.83
29.	*Nostoc* sp.	UU 62189	200	75.2	300.7	0.90
30.	*Nostoc* sp.	UU 62190	218	208.4	416.9	0.65
31.	*Nostoc* sp.	UU 512191	136	132.0	66.0	2.05
32.	*Nostoc* sp.	UU 53192	164	39.2	21.2	1.01
33.	*Nostoc* sp.	UU 513193	145	347.0	104.1	9.66
34.	*Nostoc* sp.	UU 58194	158	1980.0	990.0	0.06
35.	*Calothrix* sp.	UU 512166	188	110.0	36.6	4.54
36.	*Calothrix* sp.	UU 512167	203	832.8	557.8	2.18
37.	*Calothrix* sp.	UU 55168	161	550.0	378.1	1.24

Contd. ....

1	2	3	4	5	6	7
38.	*Calothrix* sp.	UU 518169	143	154.0	77.0	7.59
39.	*Chlorogloeopsis* sp.	UU 57178	158	991.8	270.5	0.88
40.	*Hapalosiphon* sp.	UU 512177	124	697.7	205.2	2.64
41.	*Hapalosiphon* sp.	UU 57179	146	515.6	193.4	0.74
42.	*Fischerella* sp.	UU 58159	166	1650.0	1100.0	9.21
43.	*Fischerella* sp.	UU 512180	156	790.0	230.0	10.2
44.	*Fischerella* sp.	UU 58181	137	1343.4	477.9	16.0

Serial no. 1 to 11, the organims appeared soon after wetting of the blackish brown crusts/tufts collected from various temples of India and Nepal; serial no. 12 to 44, the organisms appeared after culturing the algal crusts/tufts in BG 11 ± N medium up to 60 days.

Analysis of these results showed that a number of different species of *Tolypothrix*, few species of *Lyngbya* or *Gloeocapsopsis* occurred as epilith on the temple and monument walls of various regions of India and Nepal. There was no clear specificity of occurrence of a particular species to any region except that of a species of *Tolypothrix*, *T.byssoidea* (UU 53170) occurred on the temples of coastal regions of Orissa state and another species under the same genus i.e *T.nodosa* (UU 55175) on the monuments of Central Karnatak. However the other 9 species of blue green algae occurred widely as epilithophytes irrespective of the difference in the climatic conditions between the localities and the rock type of the temples. Of the 33 different minor component of blue green algae appeared in the crusts and tufts, maximum number of species belonging to *Nostoc*, *Calothrix*, *Chlorogloeopsis*, *Fischerella* and *Hapalosiphon* were confined to specific localities. However, members of *Gloeothece*, *Chroococcidiopsis* and *Plectonema* occurred as the associated minor component of the crusts and tufts in many temples located in widely separated geographical regions.

Almost all the 44 species/strains of epilithic blue green algae isolated from temple walls grew extreme slowly in comparision to the aquatic forms under the respective genus (Nayak *et al.*, 1996) and possessed high dark oxygen uptake rate. All those species belonging to *Tolypothrix*, *Lyngbya*, *Phormidium*, *Plectonema* or *Gloeocapsopsis* which form predominant epilithic flora on the temples possess higher respiration rate than their corresponding rate of photosynthesis. Some of the non-heterocystous forms, viz. *Gloeothece* (UU 515152), *Chroococcidiopsis* (UU 512155 and UU 57156) and *Plectonema* (UU 519165) showed nitrogenase activity under aerobic conditions. The specieses of *Tolypothrix* which were the major component of the epilithic crusts and tufts possessed lower ARA (maximum up to 3 n mol C_2H_4. μ g Chl-a^{-1}.h^{-1}); highest nitrogenase activity was measured in

Fischerella isolates which did not occur during the summer months in the crusts or tufts when extreme desiccation condition prevail on the rock surfaces.

Further, the blue green algal forms, *Tolypothrix, Lyngbya, Plectonema* or *Gloeocapsopsis* all of which possess well defined sheath layers around their trichome form dominant epilithic flora on the temple walls. Majority of associated minor components of the crusts and tufts are also ensheathed forms or produce abundant slime around their cell/trichome. There are reports that sheath and slime layers are capable of retaining moisture (Bewley, 1979; Hoffmann, 1989; Wessels and Büdel, 1995) and this appears to be the important factor in preventing damage to the epilithic blue green algal forms during dehydration under extreme environmental conditions.

5.1.2.2 Water uptake, metabolic activity and morphological features of the blue green alga *Tolypothrix byssoidea* occurring predominantly on the exposed rock surface of temples of Orissa state, India

Microscopic observation of the blackish-brown crusts from the rock surface of Lingaraj temple, Bhubaneswar and Sun temple, Konark after soaking with distilled water for about 72 h revealed brownish *Tolypothrix* filaments. Some filaments showed cells which are much broader than their length, while others possessed only external lamellate sheaths with short trichomes (Fig. 37 a). Within 15 days of cultivation of the wetted crust in BG 11 medium, fully developed *Tolypothrix filaments* showing false branching and basal heterocysts was observed (Fig. 37 b). A rapid water uptake by the crusts was observed within 30 min and was completed after about 4 hours; weight of the crust increased upto 67 per cent within 4 h of wetting (Fig. 38). Further studies showed that the blue green algae present in the crusts even during the summer months were intensely stained within 15-20 min after treatment with 2,3,5-triphenyl tetrazolium chloride (TTC) for SH groups (Tripathy and Talpasayi, 1980). Heating the crusts in dry state to 100° C for 30 min had no effect on TTC reduction but after boiling in water for 2 min, the organism became inactive to reduce TTC. The intense TTC reduction by the organism only in dry state indicates the presence of high levels of SH groups in the cells of terrestrial blue green algae. It was also observed that the photosynthetic oxygen evolution and respiratory oxygen uptake by the organism in the crust was started approximately 10 min after wetting in BG 11 medium. Photosynthesis was increased gradually up to 4 hours followed by a decrease up to 12 hours coinciding the period during which respiratory oxygen uptake was increased significantly (Fig. 39). After 24 hours of wetting, photosynthesis and the respiratory activity of the organism was stabilized at a ratio of photosynthesis to respiration almost to 0.6:1. The higher respiration rates of the crusts was possibly due to the association of heterotrophic microorganisms in the field collected samples. However, the cultures of the blue green alga isolated from the crusts also showed higher respiratory oxygen uptake rate than its corresponding rate of photosynthesis (Tripathy *et al.*, 1997).

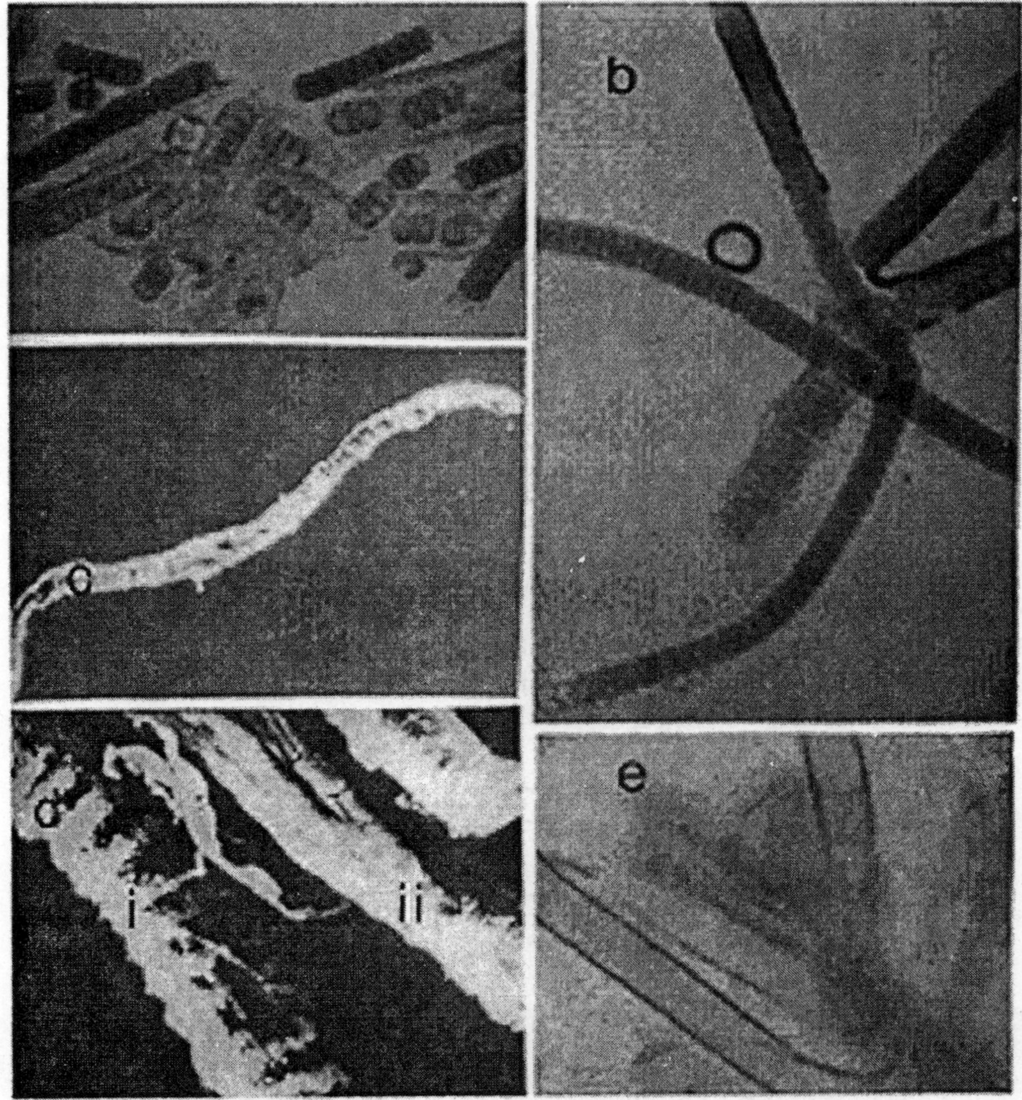

Fig. 37 (a): Blue green algal filaments containing short trichomes and/or germinating cells surrounded by brownish sheaths around the trichome appeared in culture of the crust after cultivation (x 540). (b) Light microscopic photograph of *Tolypothrix byssoidea* filaments in uni-algal culture showing trichome with heterocyst false branching and brownish sheath layer (x 540). (c) Scanning electron micrograph of a single *Tolypothirx byssoidea* filament (x 210). (d) Scanning electron micrograph of (i) a portion of *Tolypothrix byssoidea* trichome (x 1320) and (ii) the sheath after it was shed from the trichome. (e) Light microscopic photograph of isolated sheath fraction retaining the brownish pigment (x 540).

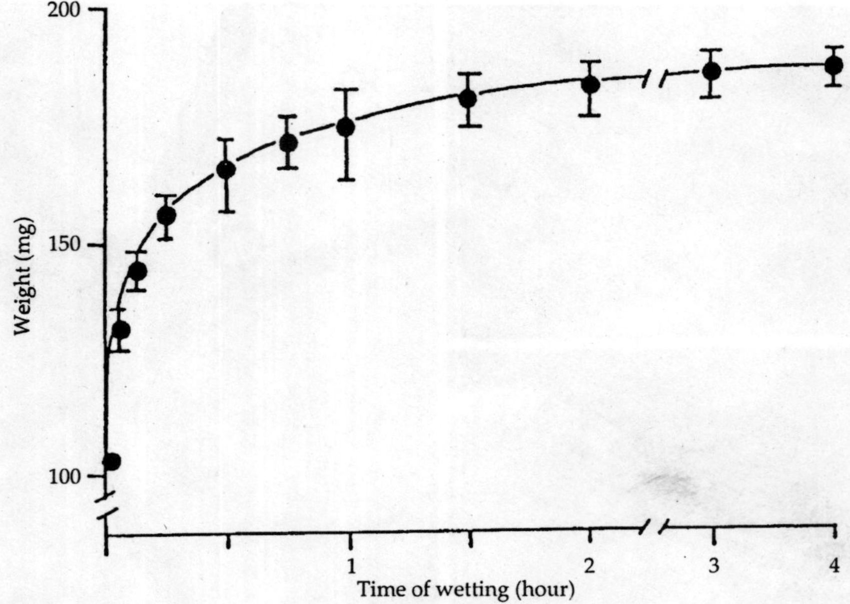

Fig.38: Time course of water uptake by the crust collected from the rock surface of Lingaraj temple.

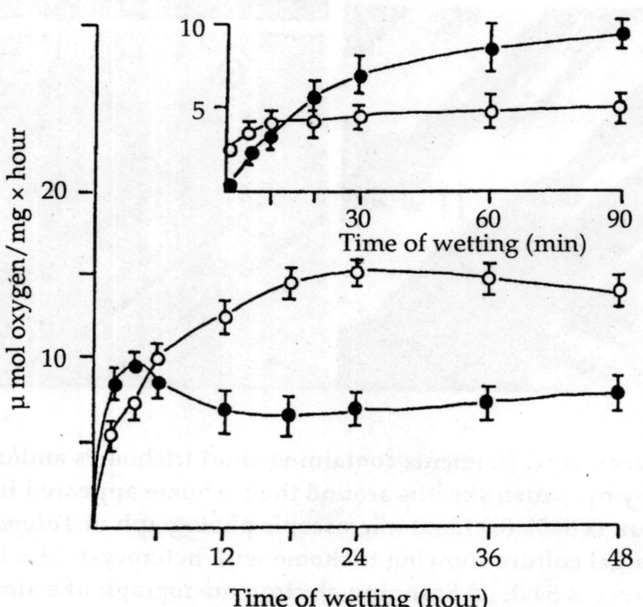

Fig.39: Photosynthesis and respiratory activity of the crust collected from the rock surface of Lingaraj temple after wetting in BG 11 medium up to 48 hours. Insert: the activity within 90 min of wetting. Photosynthesis (•), respiration (O).

These results showed that the blue green alga inhabiting the crusts on the rocks surface of the temples of the coastal region of Orissa state during summer exposed to arid condition and bright sunlight coupled with higher temperature for several months maintained reducing activity in their cells and can resume its growth on the rock surface on the onset of monsoon. Further, resumption of photosynthesis and respiration after wetting is indicative of the existing cells in desiccated state being viable even under extreme environmental conditions. Microscopic observation showed that the organisms in the crust survived in vegetative state (Fig. 37a). Further, the organism did not grow within 48 h of wetting period showing that it survived in vegetative state enduring the extreme conditions on the rock surfaces. Hess (1962) reported that members of *Scytonemataceae, Rivulariaceae* and *Nostocaceae* were resistant to dry periods of one year or longer, but not of *Oscillatoriaceae* family. Whitton *et al.* (1979) and Scherer *et al.* (1984) found that certain terrestrial *Nostoc* species survived in vegetative state without forming any endospore. Potts *et al.* (1983) demonstrated with desiccation-tolerant *Chroococcus* that internal structures remained undamaged during dryness.

Detailed morphological feature of the organism at its exponential growth phase in liquid culture was studied. The thallus became wolly and cushion-like in culture. Light microscopic photograph (Fig. 33e; 37b) and scanning electron microscopic photograph of the blue green alga (Fig. 37c,d-i) showed uniseriate filaments with occasional false branching, basal heterocyst and distinct sheath layers, shed from the filaments (Fig.37d-ii). The description of the organism is in close agreement with that of *Tolypothrix byssoidea* (Desikachary, 1959) or *Scytonema byssoideum* (C. Agardh ex Bornet et Flahault) L. Hoffmann (Hoffmann and Demoulin, 1985) with the difference that the filaments as well as the trichomes of the present organism were sometimes broader. The type of basal heterocysts has been used as the dignostic character to distinguish the genus *Tolypothrix* from *Scytonema: Tolypothrix* would have heterocysts with one pore and *Scytonema* with two pores at the base of the single false branches (Bharadwaja, 1934). Thus the accepted name for the blue green alga is *Tolypothrix byssoidea.*

T. byssoidea from rocky temple walls almost invariably showed a distinct brownish sheath. This thick sheath might allow the alga to loose or gain water from the trichomes and this appears to be an important factor in preventing damage to the organism during dehydration. The role of geletinous sheaths in retaining moisture has been emphasized (Hoffmann, 1989). It has been suggested that sheath in terrestrial algae acts as a reservoir of water, where it is bound through strong molecular forces (Campbell, 1979). These organisms react to scarcity of biologically available water by diminishing metabolic activity and the activity is quickly restored upon wetting. De Winder *et al.* (1989) attributed part of the ability of metabolic activity recovery to water retention properties of the substrate. These findings show that the sheath contributes for their survival during drought periods (Trainer, 1985).

Due to the water retention properties of the sheath, it undergoes large volume changes, collapsing when a critical amount of water is removed (Campbell, 1979). The volume changes of the sheath exert considerable force through repeated shrinking and relaxation when they are going through cycles of drying and moistening, lossening rock grains and constituting a possible factor in the gradual destruction of rocks (Jaag, 1945; Golubic, 1973; Anagnostidis *et al.*, 1983). Considerable corrosion under the microbial crusts/mats have been reported by Krumbein (1972) and Marathe and Choudhuri (1975). *Tolypothrix byssoidea* forms 1-2 mm thick crusts on the rocky walls of seven centuries old magnificent temples of Orissa state whose architetural stone carvings are disfigured or lost over the years. Most of the minute stone carvings were found corroded below the blue green algal crusts. For these reasons it can be suggested that the blue green algal foms possessing sheath layers and surviving on the rock surfaces of the temples forming characteristic crust are responsible for withering of the gigantic temples of archaeological as well as religious importance.

5.1.2.3 UV protecting pigment of the terrestrial blue green alga *Tolypothrix byssoidea*

Since *T. byssoidea* occur as crust on the rock surface of temples and thriving well at high temperature and light intensity of summer months of a tropical country like India, it was thought that the organism should have an effective protection system against intense solar radiation. In view of the seemingly decreasing stratospheric ozone layer, it is particularly interesting to study whether any protective mechanism exist in such stress tolerant blue green algae against ultraviolet radiation. Absorption spectra of methanolic extracts (90%, v/v) from equal amounts of dry blue green algal crusts and after 4 hours of wetting showed prominent absorption peaks at 384 and 330 nm, in the UV range and at 475 and 665 nm in the visible range of the spectrum (Fig. 40). Absorption maxima of the 20% (v/v) methanolic extracts of the crusts was also detected at 260 and 330 nm (Fig. 41). Absorption of the extracts from wetted crusts was higher at all these wavelengths. Prominent absorption in the ultraviolet region of the spectrum was also observed in the 20% (v/v) as well as 90% methanolic extracts of the species grown under fluorescent light (Fig. 41 and 42).

Methanolic extracts of an aquatic species of *Tolypothrix* UU 2434 isolated from rice fields and maintained in pure culture (Nayak *et al.*, 1996), which do

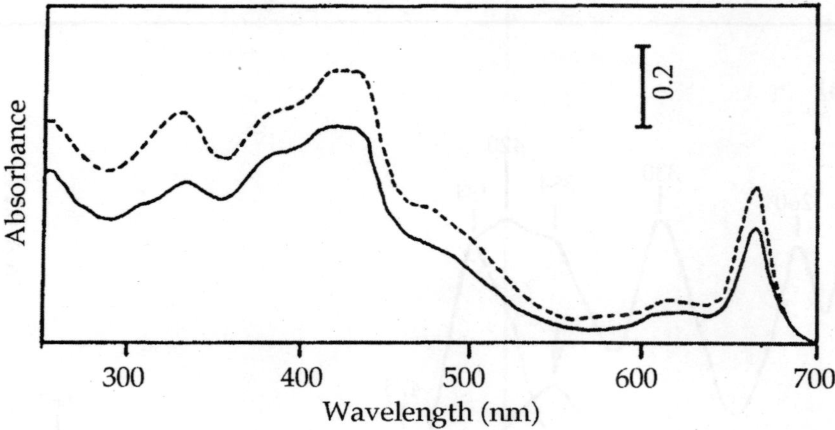

Fig.40: Absorption spectra of methanolic extracts (90%, v/v, 25° C, 3h) of the crusts from the rock surface (i) before (—) and (ii) after wetting for 24h (---). For extraction of pigments, equal amounts of the crust (air dried weight = 30 mg) were taken and extracted in 5 ml of 90% (v/v) methanol.

The absorption maxima observed at 384 and 260 nm was due to scytonemin and at 330 nm was due to the presence of mycosporine amino-acid like substances (MAAs) as observed in few other species of blue green algae (Garcia-Pichel and Castenholz, 1991, 1993; Garcia-Pichel *et al.*, 1993; Karsten and Garia-Pichel, 1996; Böhm *et al.*, 1995).

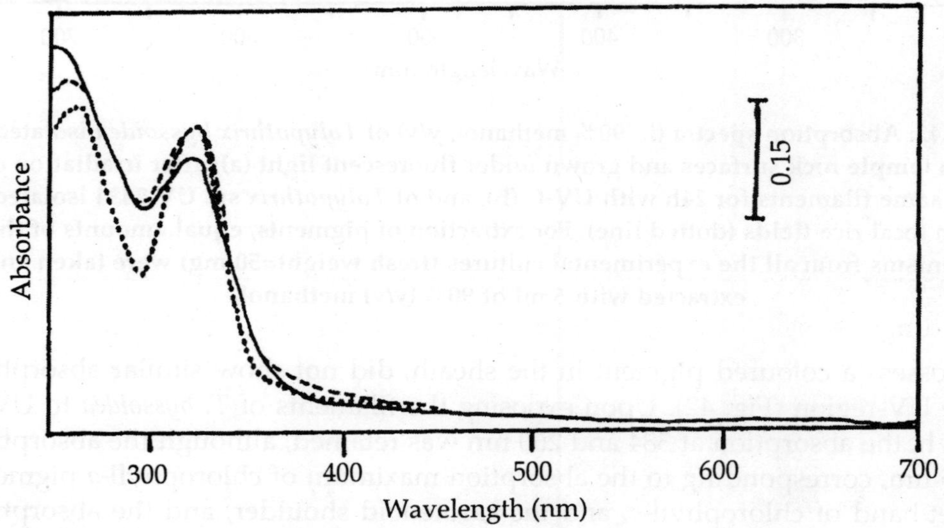

Fig.41: Absorption spectra of methanolic extracts (20%, v/v, 45° C, 2h) for mycosporine amino acid (MAA) like substances of the crusts from the rock surface (i) before (—) and after wetting for 24 h (---), and of the corresponding blue green alga isolated and grown in culture under fluorescent light (....).

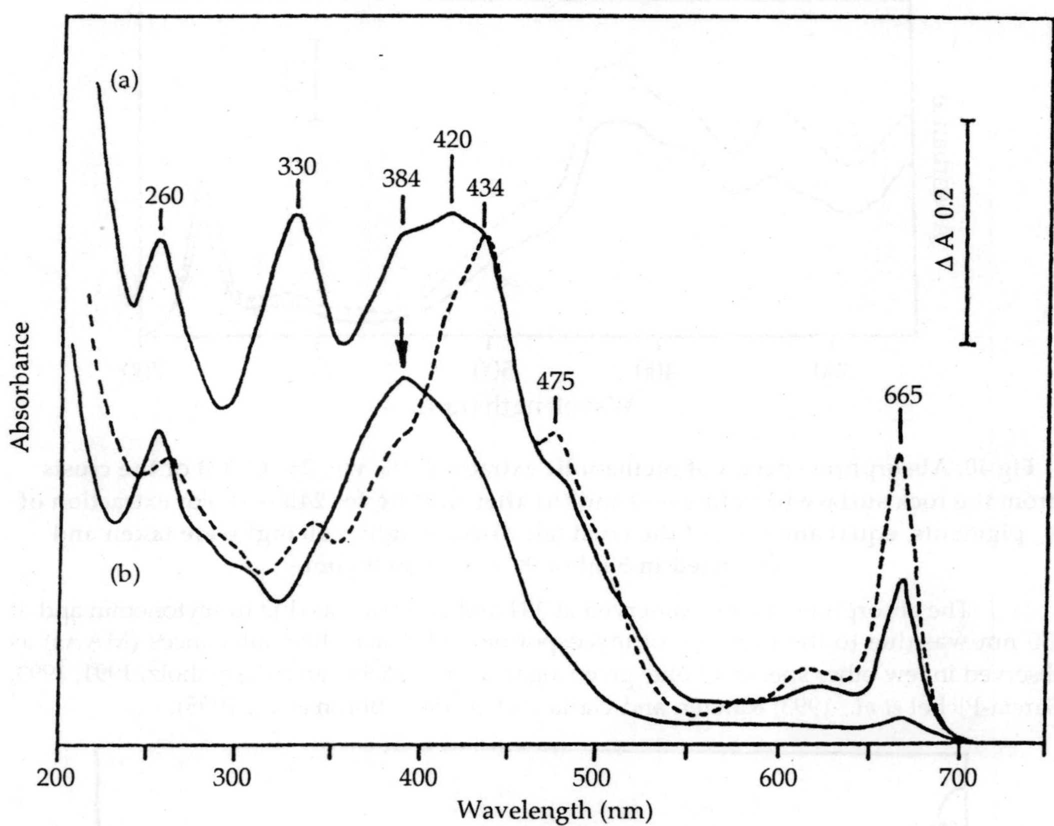

Fig.42: Absorption spectra (in 90% methanol, v/v) of *Tolypothrix byssoidea* isolated from temple rock surfaces and grown under fluorescent light (a); after irradiation of the same filaments for 24h with UV-C (b), and of *Tolypothrix* sp. UV 2434 isolated from local rice fields (dotted line). For extraction of pigments, equal.amounts of the organisms from all the experimental cultures (fresh weight=50 mg) were taken and extracted with 5 ml of 90% (v/v) methanol.

not possess a coloured pigment in the sheath, did not show similar absorption in the UV-region (Fig. 42). Upon exposing the filaments of *T. byssoidea* to UV-C for 24 h, the absorption at 384 and 260 nm was retained, although the absorption at 665 nm, corresponding to the absorption maximum of chlorophyll-*a* pigment, a soret-band of chlorophyll-*a*, and the carotenoid shoulder, and the absorption due to MAAs at 330 nm were significantly reduced (Table-26). The terrestrial species *Tolypothrix*, *T. byssoidea*, could survive even after irradiation with UV-C up to 24 h, whereas the aquatic species of *Tolypothrix* UU 2434 was killed with exposure to UV-C for 30 min (Fig. 43).

Tolypothrix byssoidea filaments collected from the trestrial habitat possessed a thick pigmented sheath that was retained when grown under various light regimes and also after exposure to prolonged UV-C radiation. Absorption spectra of the isolated sheath fraction (Fig. 37e) as well as its methanolic extracts (90%, v/v) showed prominent absorption peakes at 384 nm and 260 nm (Fig. 44). Recently, the yellow-brown pigment (scytonemin) located in the extracellular sheath of a variety of blue green algal species thriving in habitats exposed to intense solar radiation was extracted and characterized (Proteau *et al.*, 1993).

Table-26: Pigment composition (absorbance of methanolic extract at 665, 384 and 260 nm) of *Tolypothrix byssoidea* before and after UV-C irradiation.

Duration of UV-C irradiation (hour)	Absorbance of methonolic extract at				Ratio of absorbance
	665 nm	384 nm	330 nm	260 nm	665:684:330:260
0	0.15+0.02	0.46+0.07	0.47+0.05	0.46+0.06	1:3.06:3.13:3.06
1	0.15+0.03	0.46+0.06	0.38+0.06	0.46+0.04	1:3.06:2.53:3.06
3	0.12+0.02	0.46+0.04	0.26+0.08	0.46+0.03	1:3.83:2.16:3.83
6	0.09+0.02	0.44+0.03	0.14+0.04	0.42+0.04	1:4.88:1.55:4.66
12	0.07+0.01	0.41+0.06	0.06+0.01	0.35+0.03	1:5.85:1.16:5.0
18	0.04+0.02	0.37+0.05	0	0.31+0.03	1:9.25:　7.75
24	0.03+0.01	0.33+0.03	0	0.26+0.02	1:11.0:　8.66
36	0	0.26+0.04	0	0.21+0.02	
48	0	0.22+0.03	0	0.14+0.03	

Values represents the mean of five replicates ± S.D.

It shows a prominent absorption maximum in the near-UV range of the spectrum at 384 nm, which also extends up to 260 nm in the UV-C. The results also showed that methanol extracted pigments from the cells as well as from the isolated sheath fractions absorb prominently at 384 and 260 nm and were also retained after prolonged UV-C irradiation; the absorption at 330 nm due to mycosporine aminoacid-like substances was lost upon exposure to UV. These results show that the pigment present in the sheath of the terrestial blue green algae absorbs strongly in the UV-A and also in the UV-C reflecting an adaptive strategy of the organism to cope with damaging short-wavelength solar radiation.

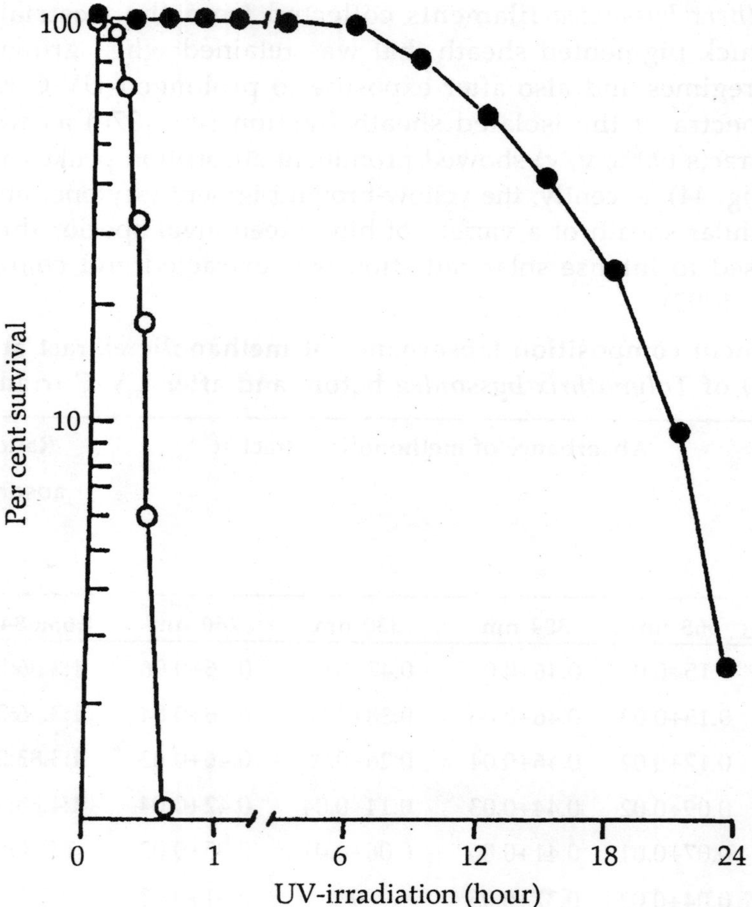

Fig.43: UV survival in two species of *Tolypothrix* isolated from different habitats. o, *Tolypothrix* sp. UU 2434 from water-logged rice fields; ● *Tolypothrix byssoidea* from temple rock surfaces.

5.1.2.4 Heat shock and radiation-stress-induced proteins in the terrestrial blue green alga *Tolypothrix byssoidea*

The cellular stress response, or heat-shock response, is involved in protecting organisms from damage due to exposure to a wide variety of stressors, including elevated temperature, ultraviolet (UV) light and xenobiotics. The stress response entails the rapid synthesis of a suite of proteins, referred to as stress proteins which are heat inducible. These proteins are highly conserved and have been found in organims as diverse as bacteria, yeast, plants, invertebrates and vertebrates including humans (Craig, 1985; Schlesinger, 1986; Margulis *et al.*, 1989). It has been suspected that stress proteins confer protection from

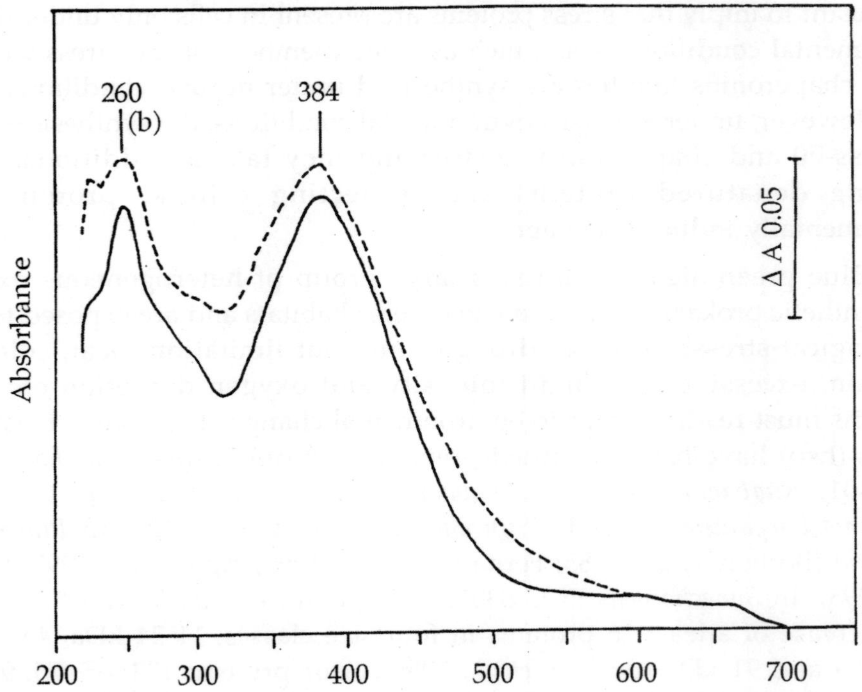

Fig.44: **Absorption spectra of the isolated sheath fraction of *Tolypothrix byssoidea* (a), and its methanolic extract (b). Fresh weight of the sheath material used for spectral analysis was 25 mg. Pigment was extracted from the same amount of isolated sheaths using 5 ml of 90% (v/v) methanol.**

environmentally induced cellular damage. The accumulation of stress proteins correlates with acquired tolerance, wherein exposure to a mild stress regime confers the ability to survive a subsequent but more severe stress that otherwise would be lethal to the organism (Lindquist, 1986).

This protections appears to Involve common targets of environmentally induced damage because tolerance is enhanced as long as stress proteins are elevated and is independent of the specific chemical or physical properties of the stressor.

Four major heat-shock proteins (hsp) families of 90, 70, 60 and 16-24 kDa are the most prominent and have been frequently referred as hsp 90, hsp 70, hsp 60, and the low molecular weight (LMW) hspS, respectively (Schlesinger and Hershko, 1988). However, recently the terms 'stres-90' and 'stress-70' are used for proteins in the 90- and 70-kDa families, respectively and that the 60-kDa family referred to as chaperonin (Gething and Sambrook, 1992). The LMW stress proteins are designated by their molecular weight. The term stress protein is not meant to imply that stress proteins are present in cells only under stressful environmental conditions is as much as some members of the stress-90, stress-70, and chaperonins families are synthesized under normal conditions (Craig, 1983). However, under adverse environmental conditions, the synthesis of stress-90, stress-70 and chaperonin increases and they take an additional role of repairing denatured proteins and protecting cellular proteins from environmentally induced damage.

Blue green algae comprise a large group of heterogeneous oxygenic photosynthetic prokaryotes that live in diverse habitats and are exposed to many physiological stresses such as drought, nutrient limitation, heat, cold, light limitation, excessive solar light intensity and oxygen depletion etc. These organisms must readily adapt to environmental changes to survive. Heat shock proteins (hsp) have been identified previously in organisms, *Synechococcus* sp. PCC 6301, *Anabaena* sp. L-31, *Anabaena torulosa*, *Anabaena* sp. PCC 7120, *Plectonema boryanum* PC 3606, *Synechococcus* sp. PCC 7942 and *Phormidium laminosum* (Borbely *et al.*, 1985; Fernandes *et al.*, 1993; Apte *et al.*, 1997; Blondin *et al.*, 1993). In *Synechococus* PCC 6301, a temperature shift from 39° C resulted in an increase of atleast 16 proteins in four size classes, 11-24 kDa, 45-49 kDa, 61-79 kDa and 91 kDa (Borbely *et al.*, 1985). Four proteins (32, 65, 75, 92 kDa) were identified when *Anabaena* L-31 cultures were moved from 30° C to 40° C and several additional stress proteins (19, 23, 82 kDa) were synthesized during heat shock and other stresses such as high salinity and osmotic stress (Bhagwat and Apte, 1989; Fernandes *et al.*, 1993). *Anabaena* sp. PCC 7120, *Plectonema boryanum* PCC 6306 and *Synechococcus* sp PCC 7942 produced 33, 35 and 19 heat shock proteins respectively though the response to heat shock was consistent for all the three strains (Blondin *et al.*, 1993). Chitnis and Nelson (1991) have isolated the *Synechocystis* PCC 6803 genes, groEL and dnaK, which encode proteins of the chaperonin 60 and 70 kDa heat shock protein families, respectively. However, the full spectrum of proteins synthesized in response to heat shock has not been identified or characterized.

When some organisms are exposed to elevated, sublethal temperatures, they acquire thermal resistance to previously lethal temperature (Marimoto *et al.*, 1990). Further studies showed that cells that do not produce functional heat shock proteins cannot become thermotloerant (Webb *et al.*, 1990), and those that over produce heat shock proteins show increased thermotolerance (Laszlo and Li, 1985; Plesset *et al.*, 1982). Although the biochemical and molecular nature of thermotolerance is not well understood, evidence suggest that heat shock proteins are required for the development of thermotolerance (Lindquist, 1988). The heat shock response of the blue green algal crust freshly collected from exposed rock surface of the temples during summer season and the corresponding organism, *Tolypothrix byssoidea* growing in culture was compared by polyacrylamide gel electrophoresis (PAGE), and the ability to develop tolerance to UV-C, gamma radiation and heat (50° C) was examined.

Upon exposure of the blue green algal crusts to UV-C radiation, none of the proteins were repressed, two high molecular weight stress proteins (HMWs) 110 and 130 kDa induced and several polypeptides of low molecular weight group (LMWs) 10.5, 19, 25 kDa; 42 kDa group (39 and 41 kDa), chaperonin (57 kDa) and stress-90 group (84 kDa) were over produced. However, when the crusts were irradiated with 100 kr gamma radiation dose or exposed to heat (50° C, 1 h), 4 or 6 polypeptides respectively belong to 42 kDa, chaperonin, stress-70, stress-90 and HMWs group were repressed and none of the proteins were neither induced nor over produced. These results show that the blue green algal crust occurring on the exposed rock surface of the temples over produce specific proteins when exposed to ultraviolet radiation, synthesized two HMWs and counteracted the UV damage. Upon exposure to gamma radiation (used in the experiments to compare the effect of ionizing and non-ionizing radiation), and heat, few polypeptides were repressed showing their adverse effect on the survival of the organism in the crust which was previously subjected to higher temperature in desiccated state with natural habitat for longer duration. However, in the control where the crust collected from nature was used, several LMWs proteins (10.5, 13, 25 kDa), Wsp 39 kDa, additional proteins under 42 kDa group (43, 49 kDa), a chaperonin (58 kDa) and stress-90 group (84 kDa) were prominent (Fig. 45; Table-27) which possibly is the reason for the adaptation of the blue green alga there-in to the extreme temperature and desiccation of the natural habitat. LMW stress proteins, composed of protein in the 20-kDa range, e.g. 10.5, 19, 25 kDa which are present in blue green alga in the crust and over produced upon exposure to UV (Table-27 and 29) are of special importance as unlike stress-90 and chaperonin, these stress proteins are not synthesized under normal conditions. Their synthesis is induced under adverse environmental conditions and had long been implicated in thermotolerance (Bosch *et al.*, 1988; Landry *et al.*, 1980; Lindquist, 1986).

The protein profile of the crusts and the corresponding blue green alga in laboratory culture showed a close similarity. However, several proteins were synthesized in the cells inhabiting the crusts, many of them being prominent (Fig. 45, 46; Table 27, 28). This may be because, the cells in the crust live in a different environment than those in the culture; in the former condition several stresses might have altered the protein synthesis of the organism. Further, the crust might not contain only *Tolypothrix* cells and proteins of several other microbes in the crust living in dormant state might have been expressed as visualized in the autoradiographs. Contrary to these results obtained in the stress

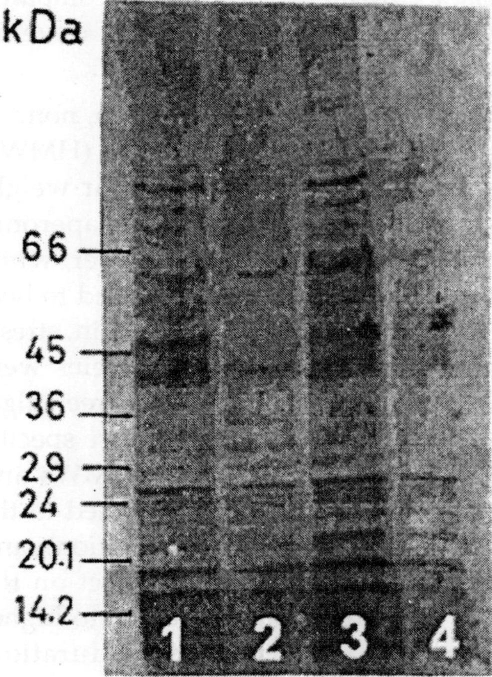

Fig.45: Stress-induced modifications of protein synthesis in blue green alga inhabiting the crusts. Equal weight of the crusts (10 mg) soaked in 1 ml of BG 11 medium and exposed to gamma radiation (lane-2), UV-C (lane-3) or heat shock (lane-4). Untreated crusts serve as control (lane-1). The crusts in BG 11 medium (1 ml) were pulse-labelled with 60 μ Ci of methionine [^{35}S] for the last 5 min under each stress. Total cellular proteins were extracted, added at equal counts per min per lane, and resolved by SDS-PAGE on a 5-14% linear gradient gel followed by antoradiography. The numbers on the left of lane-1 show the molecular masses of standard proteins in Kilodaltons.

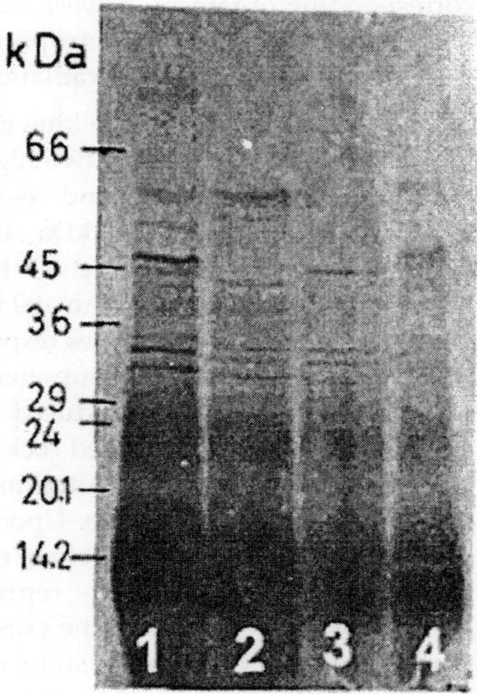

Fig.46: Stress-induced modifications of protein synthesis in in *Tolypothrix byssoidea* from culture. Equal quanity of cells of the organism were resuspended in 1 ml of BG 11 medium and exposed to gamma radiation (lane-2), UV-C (lane-3) or heat shock (lane-4). Untreated filaments of the organism serve as control (lane-1). The organism in BG 11 medium (1 ml) were pulse-labelled with 60 μ Ci of methionine[^{35}S] for the last 5 min under each stress. Total cellular proteins were extracted, added at equal counts per min per lane, and resolved by SDS-PAGE on a 5-14% linear gradient gel followed by antoradiography. The numbers on the left of lane-1 show the molecular masses of standard proteins in Kilodaltons.

response of the cells in the crust, when the *Tolypothrix byssoidea* cells grown in laboratory culture were exposed to similar UV-C dose, six proteins (28, 36, 40, 49, 52.5 and 105 kDa) belong to LMWs, 42 kDa, chaperonin and HMWs groups repressed; three (57, 84 and 134 kDa) under chaperonin, stress-90 and HMWs respectively were induced, and one 41 kDa was over produced (Table 28 and 29; Fig. 46). Similarly, on exposing *T. byssoidea* cells to 100 kr gamma radiation, 8 proteins: LMWs (28, 36 kDa), 42 kDa (43, 49 kDa), stress-96 (84, 90 kDa) and HMWs (105, 110 kDa) repressed; 3 proteins (57, 96 and 140 kDa) under chaperonin and HMWs induced; and one, 58 kDa was over produced. But when the organism from culture was exposed to heat (50° C, 1h), only three proteins of LMWs (18, 36 and 40 kDa) were repressed, two under HMWs (96 and 130 kDa) induced and two proteins, one each under 42 kDa and chaperonin (43 and 57 kDa) were over produced.

Table 27 : Protein profile of the blue green algal crust from the temple walls exposed to gamma radiation, UV-C and heat stress

Molecular weight (kDa)	Control	Stresses		
		Gamma radiation	UV-C radiation	Heat (50° C)
1	2	3	4	5
10.5	+*	+	+*	+
13	+*	+	+	+
19	+	+	+*	+
19.5	+	+	+	+
21	+	+	+	+
22.5	+	+	+	+
23	+	+	+	+
25	+*	+	+*	+
28	+	+	+	+
30	+	+	+	+
33	+	+	+*	+
39	+*	+	+	+
41	+	-	+*	+
43	+*	+	+	+

Contd. ...

1	2	3	4	5
49	+*	+	+	-
50	+	+	+	+
57	+*	+	+*	+
58	+	+	+	+
62	+	-	+	+
69	+	-	+	+
76	+	+	+	+
84	+*	+	+*	+
96	+	+	+	+
110	-	-	+	+
130	-	-	+	+
140	+	+	+	-
160	+	-	+	-*

* Prominent band in the autoradiograph showing over production of the polypeptide.

Table 28 : Protein profile of *Tolypothrix byssoidea* after exposure to gamma radiation, UV-C and heat stress

Molecular weight	Control	Stresses		
(kDa)		Gamma	UV-C	Heat
		radiation	radiation	(50° C)
1	2	3	4	5
14	+	+	+	+
16	+	+	+	+
19.5	+	+	+	+
21	+	+	+	+
22.5	+	+	+	+
23	+*	+	+	+
25	+*	+	+	+
28	+	-	-	-
32	+	+	+	+
33	+	+	+	+
37	+	-	-	-

Contd. ...

1	2	3	4	5
39	+	+	+	+
40	+	+	-	-
41	+	+	+	+
43	+	-	+*	+
49	+	-	-	+*
52.5	+	+	-	+
57	-	+	+	-
58	+	+*	+	+*
94	-	-	+	-
90	+		+	+
96	-	+	-	+
105	+	-	-	+
110	+	-	+	+
130	-	-	+	+
140	-	+	-	-

* Prominent band in the autoradiographs showing over production of the polypeptide.

Table 29 : Modification of protein synthesis in *Tolypothrix byssoidea* and the blue green algal crust from the temple walls exposed to gamma radiation, UV-C and heat stress

Organism used	Modification	Stress	Protein molecular mass (kDa)
Crust from the temples	Repression	Gamma radiation	41,62,69,160
		UV-C radiation	-
		Heat (50° C)	49,62,69,96, 140,160
	Induction	Gamma radiation	-
		UV-C radiation	110,130
		Heat (50° C)	-
	Over production	Gamma radiation	
		UV-C radiation	10.5,19,25, 39, 41, 57, 87
		Heat (50° C)	-
Tolypothrix byssoidea from	Repression	Gamma radiation	28,37,43,49,84,90, 105,110

culture	UV-C radiation	28,37,40,49, 52.5,105
	Heat (50°C)	28,37,40
Induction	Gamma radiation	57,96, 140
	UV-C radiation	57,84,130
	Heat (50°C)	96,130
Over	Gamma radiation	58
production	UV-C radiation	41
	Heat (50°C)	43,57

These results showed that unlike the modification pattern of protein synthesis in the blue green alga inhabiting the crust, the organism grown in culture could tolerate heat by synthesizing two HMWs, though three proteins under LMW group were repressed. Several proteins of diverse groups were also repressed in the organism upon exposure to UV-C and ionizing radiation stress, however, the adverse effect of repression of protein was counteracted by induction of three new polypeptides under chaperonin and HMWs group and over producing one under 42 kDa and chaperonin groups respectively.

Exposure to heat, UV-C or ionizing radiations resulted in alterations in the blue green algal protein synthesis. Three prominent types of modifications were noted; (i) synthesis of several proteins declined, (ii) synthesis of certain other proteins were selectively enhanced, and (iii) synthesis of new types of proteins was induced *de novo*. Some of these responses were, however, observed under all stress conditions (Table-29). Upon exposure to heat or radiation stress, the blue green algal cells in the crust repressed certain proteins, depending on the stressor, however, the two important water stress proteins (Wsps), e.g.33-and -39 kDa polypeptides (Scherer and Potts, 1989, Donna *et al.*, 1994) remained unchanged. Under hydrated condition (cells from culture) also these Wsps remained intact in the cells even when subjected to the stresses. The commonality of certain proteins repressed or induced in response to such divergent stresses indicates that these proteins have an important role in the maintenance of vital cellular functions of the organism even when subjected to the stresses of the extreme environment on the exposed rock surface of the temples.

5.2 SURVIVAL STRATEGIES OF DIAZOTROPHIC BLUE GREEN ALGAE OCCURRING AS MICROBIOTIC CRUST IN RICE FIELDS UNDER DROUGHT CONDITIONS

Microbiotic crusts, also known as soil algal crusts or cryptogamic crusts, cover extensive portions of the arid and semiarid regions of the world. These crusts consists of water stable surface soil aggregates held together by blue green or green algae and other microorganisms. Researchers have gathered evidence that these communities may be ecologically important as they benefit the soil on which they grow in a number of ways. The role of desert crusts is considerable as they stabilize the surface soil and consequent reduction of soil erosion increase that are otherwise highly erodible (De Winder *et al.*, 1989 a,b; Johansen, 1993). Nitrogen fixation by free living and lichenized blue green algae in the crusts add benefit to the soil. Algal crusts increase the organic carbon content of the soil they cover (Shubert and Starks, 1978) and also increase phosphorus levels in the soil by retaining fine soil particles (Kleiner and Harper, 1977).

Filamentous blue green algae frequently form a mat on the soil surface of rice fields after harvest during times of moist and moderate weather. *Microcoleus, Phormidium, Plectonema, Schizothrix, Nostoc, Tolypothrix* and *Sytonema* are the most common genera in such habitats and also in both hot and cold deserts worldwide (Ashley *et al.*, 1985; Cameron, 1960; Johnsen *et al.*, 1993). Almost all of these species possess a thick sheath layer or abundant slime around their cell wall. These ensheathed organisms represent a sizeable proportion of the microbial prototrophic soil population, especially in arid soils, where, because of their capability to withstand desiccation, high temperatures, unsuitable pH and nutrient deficiency they can surpass algal microflora (Stewart, 1980; Starks *et al.*, 1981; Skarpe and Henriksson, 1987).

In recent years the role of diazotrophic blue green algae in the nitrogen enrichment of the soil ecosystem and in the maintenance of soil fertility has become widely recognized (Whitton *et al.*, 1979; Das *et al.*, 1991). This may explain present interest in the practice of soil inoculation with nitrogen fixing blue green algae (Rodgers *et al.*, 1979; Venkataraman, 1981; Roger *et al.*, 1987; Metting, 1988). However, their fate in the soil ecosystem presents some uncertainities since, unlike the symbiotic nitrogen fixers, they lack the support of a host. For this reason, the effectiveness of soil inoculation depends primarily on the capability of the organism employed to grow, colonize and survive in the soil, especially under desiccated conditions during the summer months. The present study focusses on the survival strategies of a heterocystous blue green algae which occur prominently as crust in the fallow rice fields after harvest of the crop in the coastal regions of Orissa state, India (Adhikary and Sahu, 1998).

5.2.1 Materials and methods

Brownish crusts were collected from the soil surface of rice fields at village Maniakati in the Surada block of Orissa state, India during February, 1996 and 1997 after harvest of the kharif crop. Within few hours of wetting, algal cells could be visualized under microscope, however, culturing for up to 3 days was required for reliable identification of the species. The species which appeared in the culture was isolated, identified as *Aulosira* sp. and maintained at Utkal University assigning a strain number UU 25118. The methods followed for culturing, growth measurements, determination of photosynthetic oxygen evolution and respiratory oxygen uptake rates measurement of nitrogenese activity (acetylene reduction assay), extraction and measurement of absorption spectra of pigments has been described previously (Chapter 5.1.1). For estimation of extracelluar carbohydrates, known quantity of freshly collected dried crust was inoculated into 20 ml of BG 11 medium in a cotton stoppered sterile flask and incubated at room temperature under continuous light. At different time intervals, 1 ml aliquots of the medium was withdrawn and the extracelluar carbohydrate content was measured following Herbert *et al.* (1971). The quantity was calculated from the standard curve prepared using glucose solution within the range 10-100 μ g. ml^{-1}.

5.2.2 Results and Discussion

Brownish patches forming characteristic wolly crusts were visible on the rice field soils after harvest of kharif crop in the coastal region of India (Fig. 47a,b). Such crusts remained in the fallow rice fields through out the summer season (March-May) exposed to intense solar radiation and the algal forms regenerate from the crust soon after receiving monsoon rain. During mid-summer months, the temperature on the soil surface goes beyond 50° C, this coupled with high light intensity and extreme dryness making them inhospitable to other algal forms and microbes. The brownish crusts contained a single blue green alga, *Aulosira* sp. possessing a well marked yellowish-brown sheath layer. Unlike the cells of blue green algae with epilithic crusts on the temples, the cells of the organism in the crust on soils appeared within 4 hours of wetting (Fig. 47 d). Heterocysts appeared after 24 hours and the filamentous structure with vegetative cells and heterocysts were clearly visible after 48 hours of wetting (Fig. 41 e).

The soil crusts absorbed water rapidly and became saturated within 10 min of wetting (Fig. 47 c, Fig. 48). High respiratory oxygen uptake was observed soon after wetting which then gradually decreased up to 30 min corresponding to the time taken for initiation of photosynthesis. Soon after the crusts became water saturated, photosynthesis and respiration rate increased gradually (Fig. 49). Though the increase in the rate of respiration continued up to 6 hours of wetting and then stabilized, photosynthetic oxygen evolution continued to increase up to 48 hours of wetting and then almost stabilized maintaining a ratio of the rate of photosynthesis: respiration = 2:1 (Fig. 49).

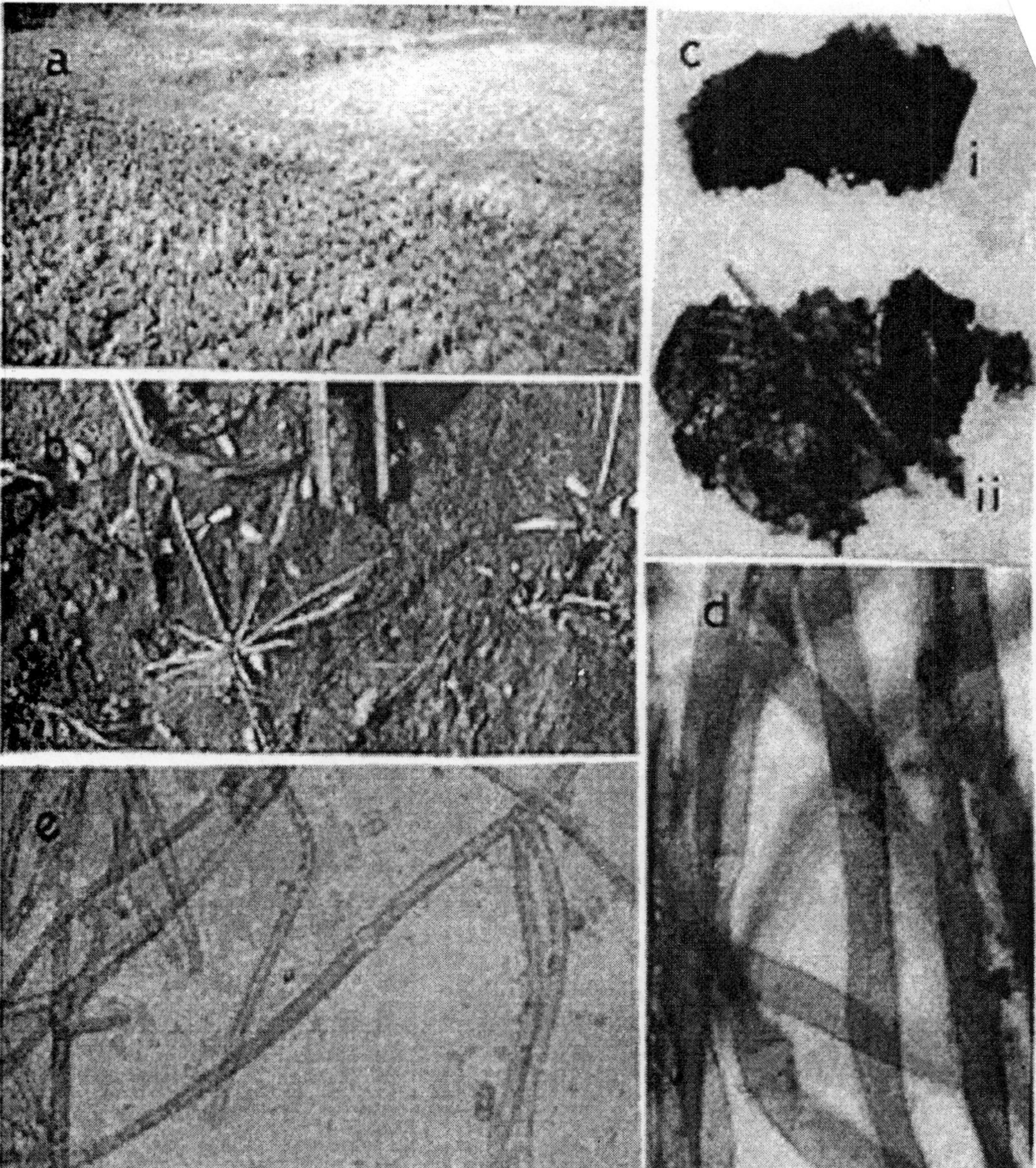

Fig. 47: (a) Blue green algal crust from the soil surface of fallow rice fields after harvest of crop; (b) blackish brown coloured crust on the soil surface of rice fields during summer season; (c) appearance of the crust before (*i*) and after wetting (*ii*); after wetting in BG 11 medium for 15 min; (d) light microscopic photograph showing well marked brownish sheath around the trichome as appeared after wetting of the crust in BG 11 medium for 4 hours (x 732); (e) appearance of trichomes, heterocyst and false branches after wetting of the crust in BG 11 medium for 48 h (x 183).

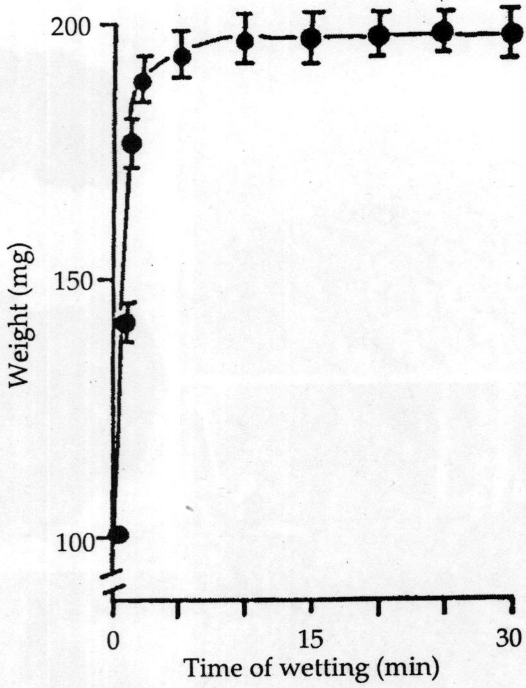

Fig.48: Time course of water uptake by the crust collected from the surface of dried soils of rice fields.

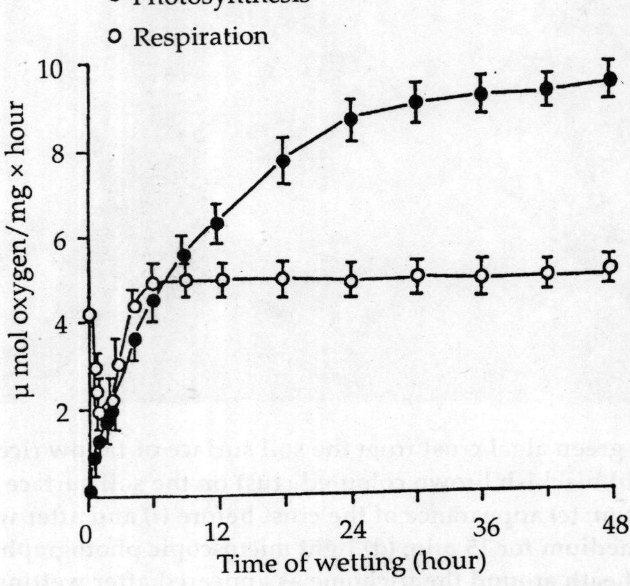

Fig.49: Photosynthesis and respiratory activity of the crust collected from the surface of dried soils of rice fields after wetting in BG 11 medium up to 48 hours.

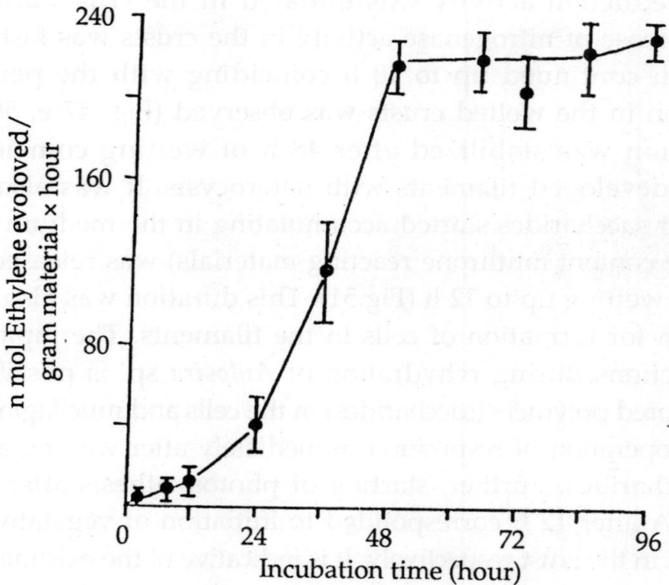

Fig.50: **Nirogenase activity (n mol ethylene evoloved per gram material per hour) of the crust from the surface of dried soils of rice fields after wetting in BG 11 medium up to 96 hours.**

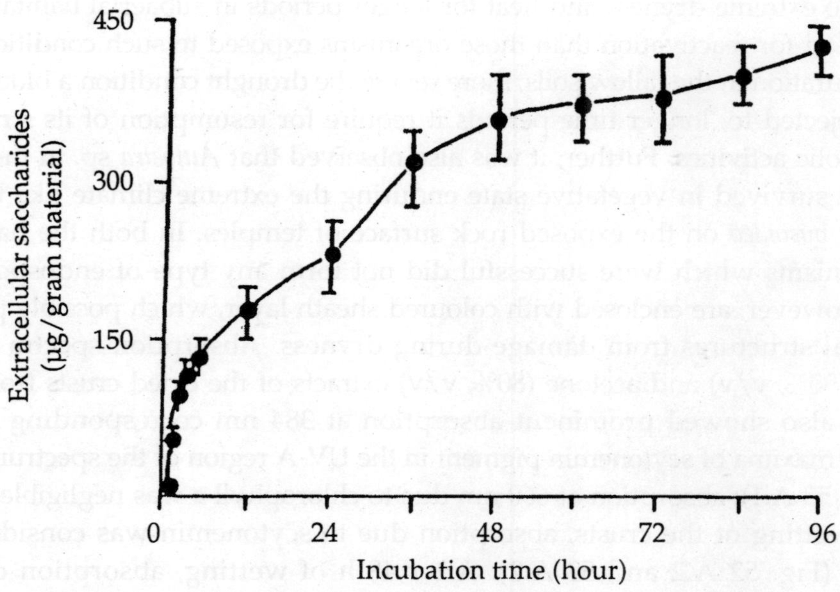

Fig.51: **Extracellular saccharide production [glucose equivalent (μ g/gram material)] by the crust from the surface of dried soils of rice fields after wetting in BG 11 medium up to 96 hours.**

Acetylene reduction activity was initiated in the crusts after 12 h of wetting period. Increase of nitrogenase activity in the crusts was faster after 24 h of wetting which continued up to 48 h coinciding with the period when heterocyst formation in the wetted crusts was observed (Fig. 47 e, 50). Rate of acetylene production was stabilized after 48 h of wetting coinciding with formation of fully developed filaments with heterocysts. It was also observed that the extracelluar saccharides started accumulating in the medium soon after wetting. Saccharide content (anthrone reacting materials) was released from the crusts quickly after wetting up to 12 h (Fig.51). This duration was also correlated with the time taken for formation of cells in the filaments. The rapid recovery of the cellular functions during rehydration of *Aulosira* sp. is possibly due to breakdown of the stored polymers (saccharides) in the cells and mucilaginous sheath layers, resulting in operation of respiration immediately after wetting and release of extracelluar saccharides. Further, starting of photosynthesis after 30 min of wetting and of ARA after 12 h corresponded to initiation of vegetative cell and heterocyst formation in the crust respectively. It is indicative of the existing vegetative cells in the crust being viable under extreme environmental conditions of the soil. However, the appearance of metabolic activities and reappearance of filamentous structure of the organism inhabiting the crust on soils was such faster than that of the species in the epilithic crusts on temple walls. This shows that the organisms subjected to extreme dryness and heat for longer periods in subaerial habitats take longer period for reactivation than those organisms exposed to such conditions for a shorter duration in the fallow soils; more severe the drought condition a blue green alga is subjected to, longer time periods it require for resumption of its structure and metabolic activities. Further, it was also observed that *Aulosira* sp. in the crust on the soil survived in vegetative state enduring the extreme climate like that of *Tolypothrix bysoidea* on the exposed rock surface of temples. In both the habitats, these organisms which were successful did not form any type of endospores or akinetes, however, are enclosed with coloured sheath layer, which possibly protect the internal structures from damage during dryness. Absorption spectra of the methanol (90%, v/v) and acetone (80%, v/v) extracts of the dried crusts from rice field soils also showed prominent absorption at 384 nm corresponding to the absorption maxima of scytonemin pigment in the UV-A region of the spectrum (Fig. 52 A.1 and 53 A.1); absorption at 660 nm due to chlorophyll-*a* was negligible. After 5 min of wetting of the crusts, absorption due to scytonemin was considerably decreased (Fig. 52 A.2 and 53 A.2). After 48 h of wetting, absorption due to scytonemin was totally reduced and the cells which were fully revived absorbed prominently at 665 and 475 nm due to synthesis of chlorophyll- *a* and carotenoid pigments. Differential spectra of the methanol (90%, v/v) and acetone (80%, v/v)

Subaerial Blue Green Algae adapted to Temperature

extracts of the crusts before wetting (Fig. 52 B.1 and 53 B.1) and after 5 m
wetting (Fig. 52 B.2 and 53 B.2) with reference to the corresponding extracts of the
wetted crusts up to 48 h showed a gradual decrease of the UV-A absorbing
scytonemin pigment proportional to the wetting time of the crusts.

These results showed that *Aulosira* sp. UU 25118 could develop coloured
sheath layers around its trichome containing UV-A absorbing scytonemin pigment
and survived in rice field soils forming crust during summer months and can resume
growth in the next cropping season when the field became waterlogged. Further,

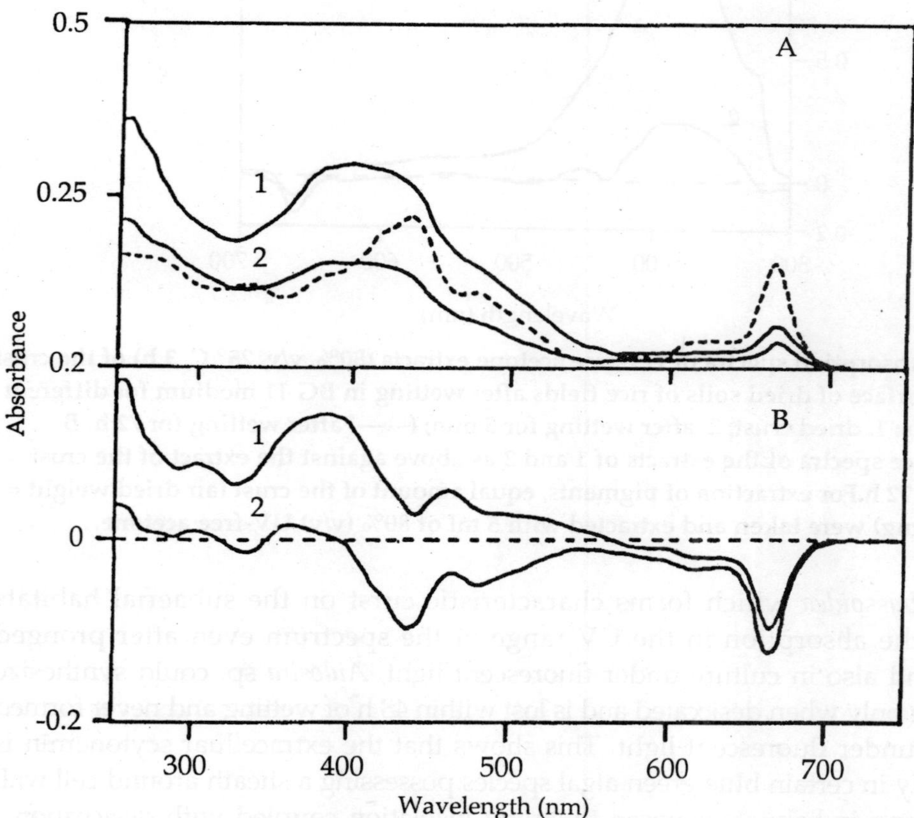

**Fig. 52: A. Absorption spectra of methanolic extracts (90%, v/v, 25° C, 3h) of the crust from
the surface of dried soils of rice fields after wetting in BG 11 medium for different
periods: 1. dried crust; 2. after wetting for 5 min; (——) after wetting for 72 h. B.
Difference spectra of the extracts of 1 and 2 as against the extract of the crust wetted for
72 h. For extraction of pigments, equal amount of the crust (air dried weight = 10 mg)
were taken and extracted with 5 ml of 90% (v/v) methanol.**

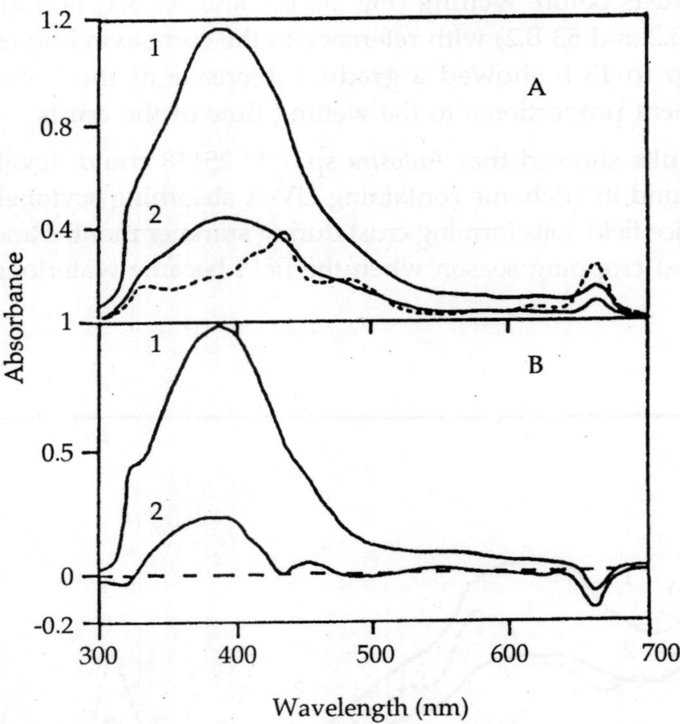

Fig. 53: A. Absorption spectra of UV-free acetone extracts (80%, v/v, 25° C, 3 h) of the crust from the surface of dried soils of rice fields after wetting in BG 11 medium for different periods: 1. dried crust; 2. after wetting for 5 min; (———) after wetting for 72 h. B. Difference spectra of the extracts of 1 and 2 as above against the extract of the crust wetted for 72 h.For extraction of pigments, equal amount of the crust (air dried weight = 10 mg) were taken and extracted with 5 ml of 80% (v/v) UV-free acetone.

unlike *T. byssoidea* which forms characteristic curst on the subaerial habitats retaining the absorption in the UV range of the spectrum even after pronged wetting, and also in culture under fluorescent light, *Aulosira* sp. could synthesize scytonemin only when desiccated and is lost within 48 h of wetting and never formed in culture under fluorescent light. This shows that the extracelluar scytonemin is formed only in certain blue green algal species possessing a sheath around cell wall and occurring in habitats exposed to strong insolation coupled with desiccation.

Blue green algal species isolated from such habitats have been found to contain one or more UV-absorbing mycosporine amino acid like compounds, and/ or extracelluar sunscreen pigment, scytonemin which has been suggested as an adaptive strategy of photoprotection against short wavelength irradiance (Garia-pichel and Castenholz, 1991, 1993). Certain filamentous blue green algae synthesized sunscreen pigments absorbing at 312 nm and at shorter wavelengths between 250 and 300 nm upon prolonged exposure to artificial solar radiation (Donkar and Hader, 1995) and to UV (Scherer *et al.*, 1988). The terrestrial blue green algae that colonize the rock surface of temples and occurred as crust on soil of fallow fields during summer months survived long periods of quiescence while simultaneously desiccated and exposed to solar radiation. Under such conditions, these organisms synthesize scytonemin and/or a variety of mycosporine amino acid (MAA)-like substances depending on the extreme conditions they are exposed to as an adaptation to reduce photodamage to their cells which must be of uttmost importance for their survival.

The observation that in dry environments many blue green algae have a layer of mucilage or a thick sheath, which are involved in the retardation of water loss form the cells, has been widely reported (Bewley, 1979; Whitton, 1987; Gantt, 1994). The expansion of a carbohydrate-enriched sheath has been reported as an ultrastructural effect of water stress on desert blue green algae (Potts *et al.*, 1983). Thick dense sheath material encasing filaments of *Nostoc commune* kept for 2 years in air-dry state has been reported (Peat *et al.*, 1988). The extended desiccation and subsequent rehydration did not lead to any structural damage to either vegetative cells or hetercysts. Thus the thick pigmented sheath layer of the epilithic terrestrial *Tolypothrix byssoidea* as well as the terrestrial *Aulosira* sp. from soils might be allowing to loose or gain water slowly in a controlled way from their trichomes in response to the environmental conditions they are subjected to, and this appears to be an important factor in preventing damage to the organisms during desiccation.

6

Blue Green Algae in Thermal Springs

Thermal springs through out the world are noted for their unusual microbial coenoses and are usually dominated by blue green algae (Anagnostidis, 1977; Castenholz, 1978). The distinguishing features of these specialized habitats are their elevated temperature, depressed dissolved oxygen level and high radioactivity (Andrews, 1991; Edmunds and Miles, 1991). The thermophilic blue green algae inhabit the hot springs or their drain ways at temperature up to 75° C (Brock, 1967; Castenholz, 1969). Since the most unique ecological feature of hot springs is the constancy of temperature and chemistry of water at the source, a thermophilic blue green algae has its optimal range of temperature available in the thermal gradient at any season together with the chemical limitations that exist exert their influence year round. This is quite unlike the environmental situations studied in earlier chapters, where great seasonal fluctuations of temperature and nutrient content occur.

It is generally assumed that blue green algae evolved in the early Precambrian and were responsible for the first significant increase in atmospheric oxygen. Most of the fossil-rich Precambrian material occur as stromatolites which could have originated in the aquatic thermal environments resemble the living members of blue green algae (Schopf, 1968). The question of whether the adaptations of modern thermophile to high temperature occurred in Precambrian or more recent times is unanswerable (Allen, 1953; Castenholz, 1973). The physical-chemical basis of stability of compounds and cellular structures at elevated temperature have been mostly studied with the *Bacillus stearothermophilus* and other thermophilic anaerobic bacteria (Lowe *et al.*, 1993). It may be fallacious to extrapolate these results to the blue green algae which are photosynthetic and form a different chemical environment. Brock (1967a) has suggested that the photosynthetic apparatus itself is the most temperature-sensitive component of blue green algae with the highest temperature tolerances up to 74° C. He was also of the opinion that the photosynthetic apparatus might have some inherent temperature limitations, since no photosynthetic organisms are found at > 73-75° C, where non-photosynthetic bacteria exist at much higher temperatures (Brock, 1967b). Theories which have been proposed to explain the phenomenon of thermophily assume that macromolecules, such as enzymes or membranes are more heat stable in thermophilic than in mesophilic organisms (Ljungdhal and Sherod, 1976; Singleton and Amelunxen, 1973; Souza *et al.*, 1974). These theories may well account for the heat stability of photosynthesis within the limit of the high temperatures of the natural habitats, since the photosynthetic apparatus consists of number of enzymes and structurally and functionally highly ordered membranes. However, survival of mesophilic blue green algal species by developing adaptive morphological features around their cells/trichome in thermal waters has not been reported. This seems that thermal springs are specialized natural habitats of great antiquity which encouraged growth of certain specific blue green algae with acquired metabolic characteristics.

Extensive reports are available on the limnology of several thermal springs of U.S.A., Europe, Japan, Israel and New Zealand (Castenholz, 1967, 1968, 1969, 1970; Brock & Brock 1966, 1967, 1968, 1969; Peary 1964, Stockner, 1967, 1968; Emoto, 1967; Kahan, 1969). Most of the investigations aimed at elucidating the upper temperature limit of life and to report on the organisms that occur in the hot springs. Through India has over three hundred hot water springs (Oldham and Oldham, 1882), our knowledge on the organisms inhabiting high temperature habitats is very meagre. Kirtikar (1886) was the first to record a thermal alga from India. Drouet (1938) described a few thermal algae during the Yale North Indian Expedition. Prasad and Srivastava (1965) and Thomas and Gonzalves (1965) have given an account of blue green algal vegetation in the thermal springs of Himachal Pradesh, Gujarat and Maharastra. Vasistha (1968), Saha *et al.* (1978), Jana (1978), Saha and Dutta Munshi (1983), Jha and Kumar (1986), Sinha and Chaubey (1986) have studied the blue green algal flora and the chemical consituents of about a hundred thermal springs distributed all over India. But the psysico-chemical nature of the thermal water and the organisms inhabiting the two thermal springs of Orissa have not been investigated by earlier workers. The present investigation was aimed at determining the physico-chemical characteristics and the flora and fauna of the hot water springs of Taptapani (Ganjam district) and Autri (Puri district) of Orissa state, India, and finding out the upper temperature limit for growth of a thermophilic species in culture and presence of adaptive features if any in the organism.

6.1 THE STUDY SITES

(i) Taptapani

This hot spring is situated at a distance of 56 kilometres (south wards) from Berhampur (19° 16′ N, 84° 53′ E) near a small village called Taptapani. It has a main tank, octangular in shape (constructed with bricks and cement by the local people) from where the mineral water and gases in the form of bubbles continuously escape (Fig. 54a, 55a). Each arm of the tank is 104" in length and are 120" apart from one another. It has a outlet through which the water over flows. In the main tank, the water level varies at different spots and its sandy bottom is full of rocks. The overflows are cemented at the bottom and are used for bathing purpose. The length, breadth and water height of 1st and 2nd over flows are 232" x 104" x 26" and 105" x 104" x 23" respectively. From the 2nd over flow water flows to the out side. It is believed that the water has therapeutic properties.

(ii) Autri

This thermal spring is situated at a distance of 43 kilometres (west wards) from Bhubaneswar (20° 12′ N, 85° 22′ E) near a small village called Baghamari. It has a circular main tank of 161" diameter and 168" depth (artificially constructed) from where water and gases escape from the bottom in the form of bubbles (Fig. 54 b, 55b). Similar to Taptapani it has a rocky bottom but the water depth is very deep (139"). Just above the water level there are two separate outlets through which water flows to the two separate cemented bathing tanks (overflows) (Fig. 55d). The 1st and 2nd overflows are placed at 77" and 334" distance from the main tank with an area of 135"x 135" and 120" x 120" respectively. Both the overflow tanks have around 40" of water height through out the year and from these tanks water flow to the surrounding rice-fields. During 1990 few more tanks were constructed at the site and are used for bathing of tourists (Fig. 56).

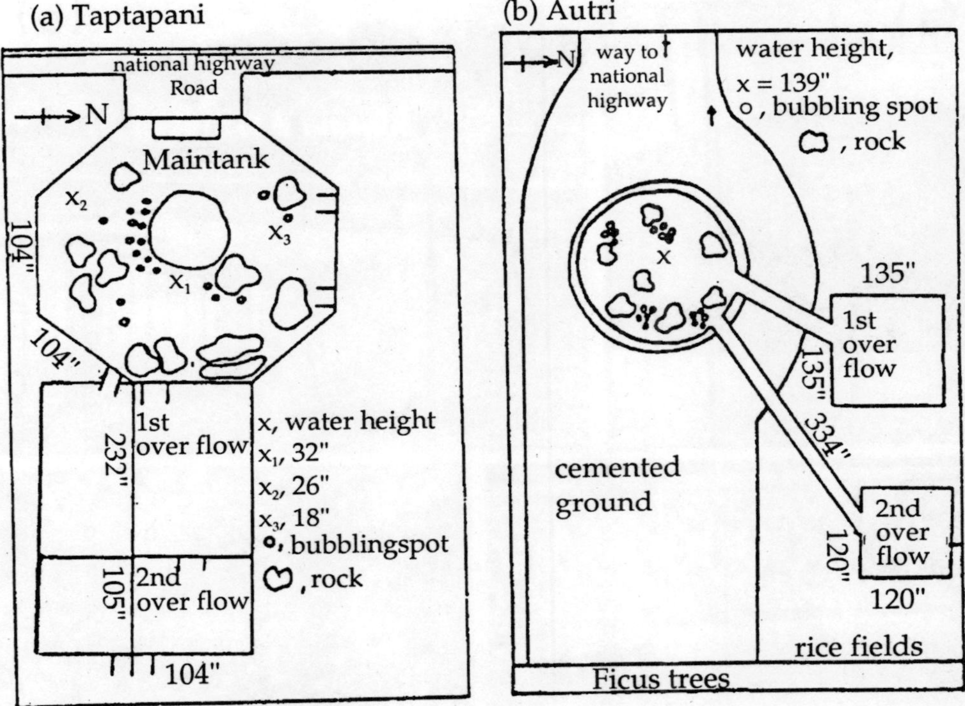

Fig. 54: Diagramatic representation of the study site of the hot water spring at Taptapani (a) and Autri (up to the year 1989) (b) in the Orissa state.

6.2 Materials and methods

Water samples and algal mats were collected from the main tank and the overflows of Taptapani and Autri between July 1981 and May 1982 at regular intervals. Temperature was recorded on the spot with a mercury thermometer graduated up to 100° C. The hydrogen-ion-concentration of the thermal waters was determined by using a digital pH meter. Water samples were analysed for the presence of various chemicals according to the standard methods for the examination of water and waste water A.P.H.A. (1965) and were indicated in parts per million. Plankton samples were collected by filtering known quantity of the water taken from different spots of the main tank and the over flows of the thermal springs through a plankton net made of standard bolting silk cloth (No. 21 with 77 mesh/sq.cm). The concentrated plankton sample was preserved in 4 per cent formalin and quantitatively determined by the sedimentation and drop count method. The various phytoplanktons were identified according to Desikachary (1954) and Fritsch (1939).

Fig. 55: (a) Main tank of the hot spring Autri (Khurda, Orissa) showing mouth of the pipe through which hot water is carried away to the overflows (bathing tanks). Very few phytoplankton occurred at this spot. (b) Bathing tanks of Autri hot spring located 200 meters away from the main tank where algae growth was conspicuous. (c) Main tank of the hot spring Taptapani (Ganjam, Orissa) showing luxurious growth of blue-green algae and the mouth of the pipe through which hot water is carried away to the over flows (bathing tanks). (d) Thick scum of the blue green algae, *Mastigocladus laminosus* Cohn in the over flows of the Autri hot spring at 48° C.

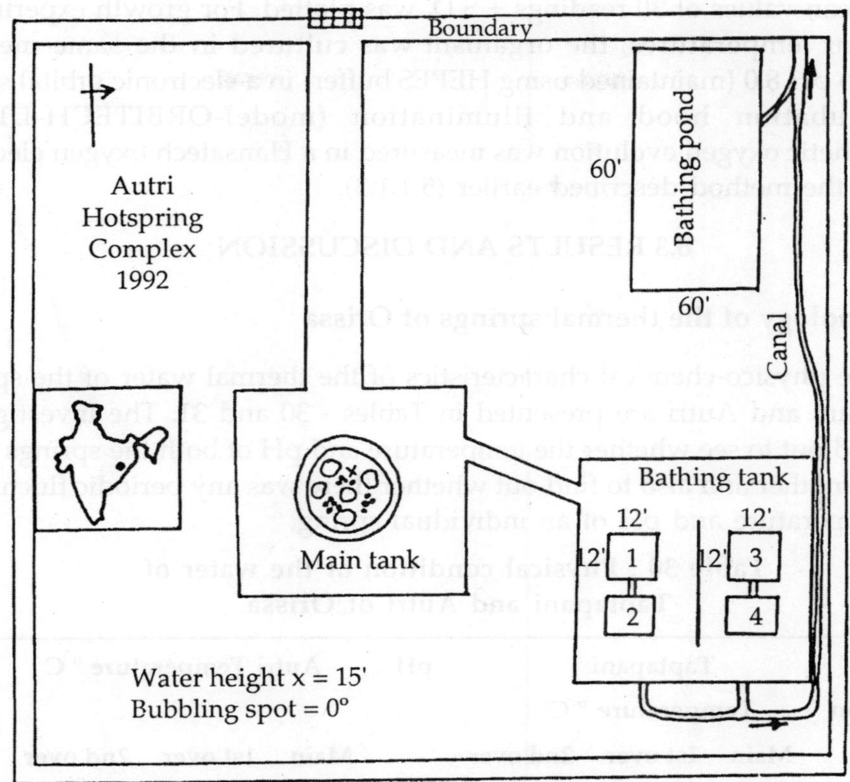

Fig. 56: Diagramatic representation of the study site at Autri with reconstructed bathing tanks for tourists ad pilgrims since 1990.

The blue green alga, *Mastigocladus laminosus* was isolated and the unialgal culture of the organism was obtained by plating methods. Axenic culture was obtained following the method described earlier (Chapter-2). The organism was grown in Allen and Arnon's medium (1955) with trace elements as used by Fogg (1949) at 30±1° C (by repeated cultivation, the organism gradually adapted to grow at this lower temperature in comparison to its natural habitat) under continuous light from day light fluorescent tubes at an intensity of 7.5 W/m^2. Growth experiments at different pH levels of the culture were performed in 100 ml Erlenmeyer flasks containing 25 ml of nitrogen-free medium. The required pH of the culture media was obtained and adjusted from time to time by aseptic addition of a few drops of 0.1N NaOH or 0.1N HCl. An equal amounts of exponentially growing organism (equivalent to 1 mg dry weight) was inoculated to the experimental flasks and the cultures were harvested after 15 days of incubation. Growth was estimated on dry weight basis. Frequency of branching was calculated by counting the number of cells at intervals between branch initials; mean values of 30 readings ± S.D. was plotted. For growth experiments at different temperatures, the organism was cultured in the same medium adjusted to pH 8.0 (maintained using HEPES buffer) in a electronic orbital shaker with incubation hood and illumination (model-ORBITECH-LT-1L). Photosynthetic oxygen evolution was measured in a Hansatech oxygen electrode following the method described earlier (5.1.1.3).

6.3 RESULTS AND DISCUSSION

6.3.1 Limnology of the thermal springs of Orissa

The physico-chemical characteristics of the thermal water of the springs of Taptapani and Autri are presented in Tables - 30 and 31. The investigation was carried out to see whether the temperature and pH of both the springs differ from one another and also to find out whether there was any periodic fluctuation of the temperature and pH of an individual spring.

Table 30 : Physical condition of the water of Taptapani and Autri of Orissa.

Date of investigation	Taptapani Temperature ° C			pH	Autri Temperature ° C			pH
	Main tank	1st over flow	2nd over flow		Main tank	1st over flow	2nd over flow	
July 1981	44.5	43.5-44.5	40.0-42.0	9.1	56.5	46.5-48.0	46.0-48.0	9.6
Sept. 1981	44.0	43.0-44.0	40.0-41.0	9.2	55.5	45.0-48.0	45.-48.0	9.4
Nov. 1981	44.0	42.0-43.5	40.0-41.0	9.2	55.0	45.0-48.0	45.0-47.0	9.7
Jan. 1982	44.0	42.0-44.0	40.0-41.0	9.2	55.0	46.0-48.0	46.0-48.0	9.7
March 1982	44.0	43.44.0	40.0-41.0	9.2	55.5	46.0-49.0	46.0-49.0	9.8

Table 31 : Chemical characteristic of the water from Taptapani and Autrı of Orissa.

Constituents	Water from the main tank of Taptapani	Water from the main tank of Autri
Free CO_2	Nil	Nil
Total solids (ppm)	580	240
Total hardness ($CaCO_3$) (ppm)	9.8	14.0
Chloride (Cl^-) (ppm)	92.0	130.0

Nitrate (NO$_3^-$) (ppm)	trace	Trace
Nitrite (NO$_2^-$) (ppm)	trace	Trace
Phosphate (PO$_4^-$) (ppm)	trace	Trace
Silicate (SiO$_3^-$) (ppm)	48.0	33.6
Dissolved O$_2$ (ppm)	4.9	1.4
Sulphurated hydrogen (as H$_2$S)	Present	Present

From the results (Table-30) it was observed that the periodic fluctuation of the temperature of both the springs was insignificant and the minor variation may be due to the change of climatic temperature of the regions during different seasons. The local inhabitants stated that the temperature of these thermal springs has remained constant over the past few decades and the location of the fissures have also remained unchanged. There were also reports that many hot springs were constant in their thermal and hydrologic properties over a few hundred years (Brock and Brock, 1967; Stockner, 1968; Castenholz, 1969). The slightly low pH of the water during July may be due to the inflow of rain water from the surrounding area. It has been analysed that the water of both the hot springs were free of CO$_2$. There were reports (Brock and Brock, 1966) that light and CO$_2$ were not the limiting factors for phytoplankton production in alkaline hot springs. The water sample of both the thermal springs investigated had traces of nitrate, nitrite and phosphorus (Table 31). However chloride, carbonate and silicate were present at various proportions which normally is not found in fresh water pools. In addition sulphurated hydrogen was present in the spring water, which emit H$_2$S gas smell. Total solids and dissolved oxygen of the water of Taptapani was more in comparison to the thermal water of Autri (Table 31). These physicochemical characteristics of the hot water springs were mainly responsible for the growth of various organisms in the spring water.

The thermal water of Taptapani encouraged the growth of a number of organisms mostly phytoplankton which imparts deep green colouration to the spring. Quantitatively highest plankton population was observed in the main tank of Taptapani. In its overflows, the plankton population was a little less and in the 2nd over flow the number of planktons was reduced. Since both the overflows are normally used for bathing, the mats were partially cleared by the tourists. The high temperature of the clear water of Autri do not encourage the growth of a large number of plankton. The over flows, where the temperature of the water was less, there was luxurious growth of various organisms. (Fig. 55c). Zooplankton were totally absent in the main tank of Autri. This may be due to the higher temperature of the spring. However, abundance of phytoplankton and zooplankton were noticed in both of its over flows. Due to higher water depth of the over flows of Autri, the tourists normally do not enter the tanks for bathing, thus do not disturb the growth of the plankton mats. The plankton mats collected from various spots of the main tank and over flows of both the thermal springs composed of mostly members of cyanophyceae/ bacillariophyceae and a few zooplankton (Table-32). Similar gelatinous and calcareous mats of various colour consisting of blue green algal cells have also been reported in other hot springs of Europe and America (Castenholz 1969; Stockner, 1967). From these results it seems that the temperature may be the major factor in determining the qualitative and quantitative distribution of planktonic organisms. The differences in the floristic patern and the productivity of the hot springs may be due to the difference in concentration of the mineral elements of the spring water.

Table 32 : Distribution of plankton in the main tank and over flows of Taptapani and Autri of Orissa

	Taptapani			Autri		
	Main tank	1st over flow	2nd over flow	Main tank	1st over flow	2nd over flow
Total plankton/ 1000 ml	438,080	322,650	150,470	650	290,200	248,560
Total zooplankton/ 1000 ml	14,150	10,820	4,400	Nil	18,600	22,350
Zooplankton percentage	3.23	3.35	2.92		6.4	8.99
Total phytoplankton /1000 ml	423,930	311,830	146,070	650	271,600	226,210
Phytoplankton percentage	96.76	96.64	97.07	100	93.59	91.0

Of the various thermophilic organisms, only *Chroococcus minor* (Nätz) Näg., *Oscillatoria teribriformis* (Ag.), *Spirulina* sp., *Mastigocladus laminosus* (Cohn), *Cyclotella* sp. and *Navicula* sp. occur in both the hot water springs and in addition, *Synechococcus lividus* (Copeland) and *Synechococcus elongatus* (Näg.) occur abundantly in Taptapani. A number of blue green algal forms, the green algae, *Euglena* and various zooplanktons were also observed in the thermal waters (Table-33). Certain thermophilic organisms, viz. *Oscillatoria teribriformis* (Ag.), *Mastigocladus laminosus* (Cohn), *Cyclotella* sp. and *Navicula* sp. occur abundantly in the thermal water even at 55° C in the hot water spring of Autri through out the year of investigation. There are records that *Spirulina* sp., *Chroococcus* sp., *Aphanothece* sp., *Anabena* sp., *Oscillatoria* sp., *Navicula* sp. and *Cyclotella* sp. have been collected from thermal springs with 26-50° temperature range (Gonzalves, 1947; Prasad and Srivastava, 1965; Vasistha 1968). *Oscillatoria* sp. has the ability to grow at high temperature in various thermal springs of India (Gonzalves, 1947; Prasad and Srivastava, 1965). In this report, the occurrence of a common thermophilic blue green alga *Mastigocladus laminosus* (Cohn) (Castenholz, 1970) at temperatures up to 55° C in the thermal water of Taptapani and Autri of Orissa is a new record from Indian thermal springs.

6.3.2 Growth and branching development of *Mastigocladus laminosus* at different pH levels

A branched blue green alga *Mastigocladus laminosus* found growing in the main tank and over flows of Taptapani hot spring displayed variable branching frequencies. The yearly mean temperature and pH of the spring water was 44° C and 9.0 respectively. Temperature and hydrogen-ion-concentration, the important factors of hot spring water, appears to control the degree of branching of this organism. Experiments were designed to find out the relation between growth and branching development of the *Mastigocladus* strain at different temperature and pH levels of the culture in the laboratory.

Table 33 : Occurrence of various organisms in the main tank and the over flows of Taptapani and Autri of Orissa.

Organism	Taptapani			Autri		
	Main tank	1st over flow	2nd over flow	Main tank	1st over flow	2nd over flow
Synechococcus lividus Copeland	++	++	++	-	-	-
Synechococcus elongatus Näg	++	++	++	-	-	-
Synechosystis aquatilis Sanv.	+	+	+	-	-	-
Aphanothece sp.	+	+	+	-	-	-
Chroococcus minor (Kütz) Näg	++	++	++	-	++	++

Organism						
Oscillatoria teribriformis Ag.	++(+)	++	++	++(+)	++(+)	++(+)
Oscillatoria princeps Vauch.	-	-	-	-	+	+
Oscillatoria tenuis Ag.	-	-	-	-	+	+
Phormidium purpursascens (Kütz) Gom.	+	+	+	-	+	+
Phormidium sp.	-	-	-	-	+	+
Lyngbya sp.	+	+	+	-	+	+
Spirulina sp.	++	++	++	-	++	++
Anabaena sp.	+	+	+	-	+	+
Mastigocladus laminosus Cohn	++(+)	++	++	++(+)	++(+)	++(+)
Scenedesmus sp.	-	-	-	-	+	+
Cosmarium sp.	+	+	+	-	+	+
Cyclotella sp.	++(+)	++	++	++(+)	++(+)	++(+)
Navicula sp.	++(+)	++	++	++(+)	++(+)	++(+)
Euglena sp.	+	+	+	-	+	+
Tobrilus sp.	+	+	+	-	-	-
Cyclops sp.	+	+	+	-	+	+
Lacane sp.	-	-	-	-	+	+
Nauplius larva	+	+	+	-	-	-

+ = Present; - = absent; ++ = Occur abundantly; (+) = Occur throughout the year.

Results on the growth and branching development of *M. laminosus* at different pH levels of the cultures (5, 6, 7, 5, 8, 8, 5, 9, 9, 5, 10, 11) is given in the Fig. 57. No growth was observed at the pH level of 5. Brock (1973) also reported that blue green algae were absent from a wide variety of acidic environments and in enriched cultures at less than pH 5. The rate of growth of *Mastigocladus* was not substantial at pH 6. Further increase in pH of the media encouraged the growth of the organism and a maximum was obtained at pH 8.5.

Growth of the alga decreased in the alkaline range at the level of more than pH 8.5 of the culture medium (Fig. 57). Microscopic examination revealed that the filaments were devoid of branches in the acidic pH. Distinct branching was observed in the alkaline pH and pH 9.5 was found to be the most effective one where the frequency of branching was observed as the maximum. Each alternate cell or even each cell of the filament grown at pH 9.5 was found to

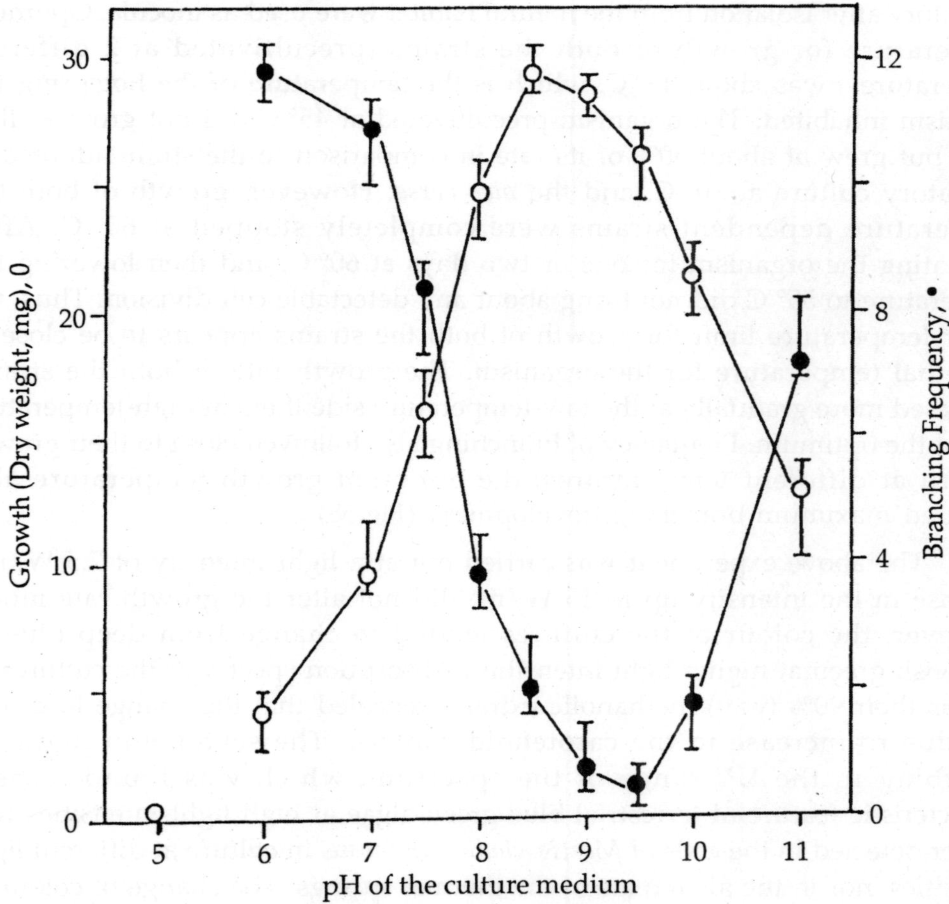

Fig. 57: Growth and branching frequency of *Mastigocladus laminosus* Cohn at different pH levels of the culture. Cultures incubated at 30° C for 15 days.

develop a branch initial. The pH which was most effective for the best growth of the organism was found ineffective at the same level for the highest degree of branching development, thus suggesting an independent role of hydrogen-ion-concentration for the branching initiation. It appears that the degree of branching is rather a function of hydrogen-ion-concentration of the growth medium and to an appreciable extent it is independent of the growth rate of the organism.

6.3.3 Temperature dependency of the growth of *Mastigodadus laminosus*

Growth and branching frequencies of the alga *Mastigodadus laminosus* at different temperature in 5 days of culture (pH, 8.0; light intensity 7.5 W/m^2) are given in Fig. 58. The organism maintained at 45° C and at 30° C in the laboratory after isolation from the natural habitat were used as inocula. Optimum temperature for growth of both the strains (precultivated at 2 different temperatures) was about 45° C, which is the temperature of the hotspring the organism inhabited. The organism precultivated at 45° C did not grow well at 30° C but grew at about 50% of its rate in comparison to the strain adopted to laboratory culture at 30° C, and the *vice versa*. However, growth of both the temperature dependent strains were completely stopped at 60° C. After cultivating the organism for one or two days at 60° C, and then lowering the temperature to 55° C did not bring about any detectable cell division. Thus, the upper temperature limit for growth of both the strains appears to be close to the lethal temperature for the organism. The growth rate of both the strains decreased more gradually at the low-temperature side them at high-temperature side of the optimum. Frequency of branching also followed close to their growth pattern at different temperatures; the optimum growth temperature also favoured maximum branching development (Fig.58).

The above experiment was carried out at a light intensity of 7.5 W/m^2. Increase in the intensity up to 15 W/m^2 did not alter the growth rate much. However, the colour of the cultures tended to change from deep blue to yellowish green at higher light intensities. Absorption spectra of the cultures as well as their 90% (v/v) methanolic extracts revealed that the change in colour was due to increase in the carotenoid content. The scytonenin pigment, absorbing in the UV-range of the spectrum, which was found to be a characteristic feature of terrestrial blue green algae at high light intensities was neither detected in the cells of *Mastigocladus laminosus* in culture at different light intensities, nor in the algal mats of the thermal springs. The change of colour of the thermophilic blue green algae in response to higher light intensities was not unusual as similar phenomenon has been reported with mesophilic species (Brown and Richardson, 1968).

6.3.3.1 Photosynthetic activities of *Mastigocladus laminosus* at different temperatures

The results on the changes in the rates of photosynthetic oxygen evolution of *M. laminous* at different temperatures shown in Fig. 59 showed that growth could be well correlated with the activity of photosynthesis over a wide range of temperature. Rates of photosynthetic oxygen evolution was high at 45° C, and decreasing as the temperature was raised above, or lowered below the optimum range. Further, the temperature profile of photosynthesis in the cells of the organism precultivated at the temperature of the natural habitat (45° C) and at ambient (30° C), was similar to that of their respective growth pattern (Fig. 58 and 59). This shows that photosynthesis in an important factor

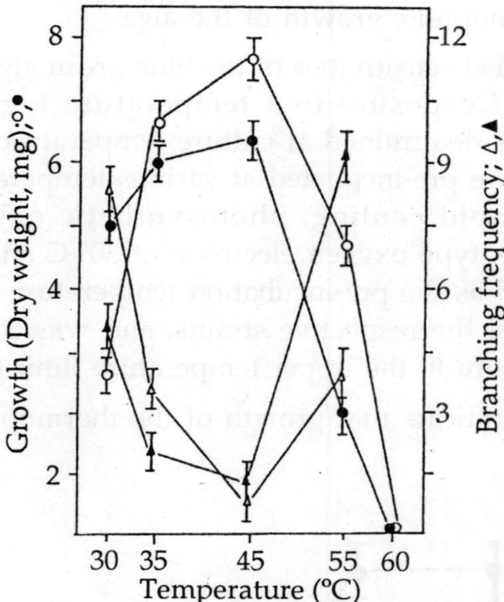

Fig.58: Growth and branching frequency of *Mastigocladus laminosus* at different temperatures. Cultures incubated for 5 days. The inoculum contained equal quantity of *Mastigocladus* cells grown previously at 30° C (closed, •) or 45° C (open, o) in culture after isolation; o• growth, ΔΔ branching frequency.

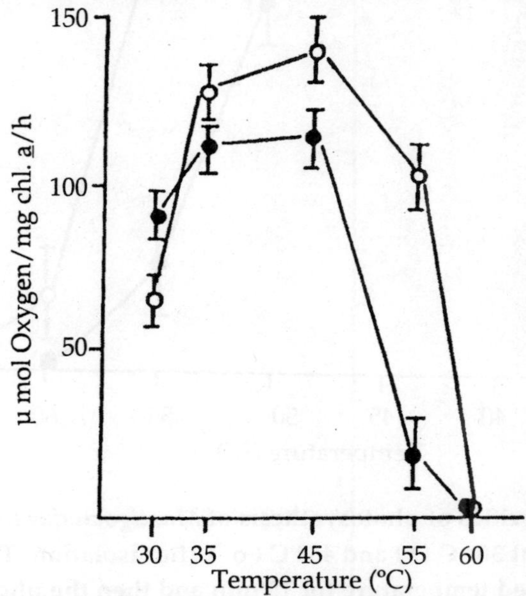

Fig. 59: Photosynthetic oxygen evolution (μ mol O_2 per mg chlorophyll-*a* per hour) of *Mastigocladus laminosus* at different temperatures. The cells grown previously in culture at 30° C (•) and 45° C (o) after isolation were used.

determining the thermophilic growth of the alga.

To study the heat sensitivities of the blue green algae, its photosynthetic activity after a brief exposure to a temperature higher than the upper temperature limit was determined at culture temperature. The cells cultivated at 30° C and 45° C were pre-incubated at various temperatures up to 65° C for 10 min, and after rapid cooling, photosynthetic oxygen evolution was determined in a Clark type oxygen electrode at 30° C and 45° C, respectively. The activity decreased as the pre-incubation temperature was raised above the culture temperature of the respective strains, and was completely inactivated even after brief exposure to the upper temperature limit (Fig. 60).

These results indicate that growth of the thermophilic blue green alga,

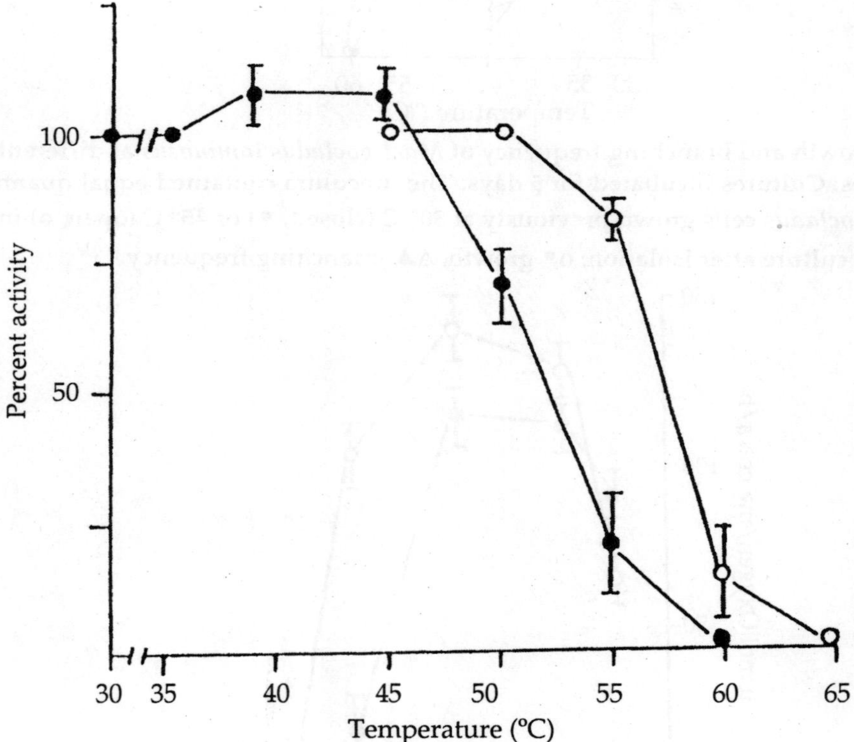

Fig.60: Heat sensitivities of photosynthesis of *Mastigocladus laminosus* grown previously in culture at 30° C (•) and 45° C (o) after isolation. The cells were pre-incubated at the indicated temperature for 15 min and then the photosynthetic oxygen evolution was measured at 30° C (•), and 45° C (o) respectively. Control (100%) activities determined were 88±9 and 137±12 µ mol O$_2$ evoloved per mg chlorophyll-*a* per hour for cells previously grown at 30° C and 45° C respectively.

Mastigocladus laminosus in closely related to the activity of photosynthesis over a wide range of temperatures. Growth was optimum at the temperature where photosynthesis was most active. Low growth rates at suboptimal temperatures could be correlated with the decline in photosynthetic activity. Higher temperatures, which brought about an irreversible inactivation of photosynthesis, were lethal to the cells.

Most of the thermophilic blue green algae grow poorly or not at all below 30 to 35° C. Certain races of *Synechococcus* of thermal springs did not grow below 50° C; others ceased growth between 50 and 40° C, some between 40 and 35 and other below 30° C (Peary and Castenholz, 1964). High-temperature clones of *Mastigocladus* from Iceland, however are able to grow over the range of 62 to 29° C and perhaps lower (Castenholz, 1969). Upper temperature limits of these organisms are often a few degrees above the optimum. The lower limit is flexible, depending on previous conditioning, light intensity, and other factors.

A number of natural habitats of high temperature exist ranging from Sun-heated soils, exposed rock surfaces in the tropics and building terraces with temperatures of 60 to 70° C, to erupting volcanos reaching 1,000° C. A question of considerable interest centres on the upper temperature limit for life. As noted by Brock (1967), "bacteria are able to grow at any temperature at which there is liquid water, even in pools which are above the boiling point". However, photosynthetic blue green algae are known to grow at constant temperatures as high as 73 to 74° C in thermal springs. A maximal growth rate at temperatures over 45° C in mainly a characteristic of procaryotic organisms. Only a few species of eukaryotic protists or animals tolerate temperatures above this (Castenholz, 1969). Thus many species of blue green algae of non-thermal habitats have higher temperature optima than the eukaryotic algae of the same environments.

Blue green algae are becoming more conspicuous in this age of increasing environmental pollution due to industries, domestic sewage and agrochemicals used in agriculture and forestry. In nutrient-enriched wastes, algal blooms are more frequent, denser, and longer lasting. The thermal pollution from the water coolant of power plants (nuclear and conventional) has, one of its most striking effects, the promotion of blue green algal growth. Thus it is probable that habitats suitable for thermophilic organisms are going to increase substantially. At elevated temperature coupled with presence of specific nutrients, certain species of blue green algae are successful due to their capability to form special adaptive morphological features such as sheath around their cell wall which protect them from the harmful effect of the surrounding environment. In the previous chapters it has been shown that extracelluar sheath provide protection to the blue green algal cells occurring in industrial waste water polluted habitats, pesticide burdened soils and also to terrestrial species subjected to periodic temperature and desiccation stress. In the thermal springs, however, blue green algal forms with a distinct sheath seldom occur, and, interestingly the maximum upper temperature limit of 74° C in such habitats was accorded in a *Synechococcus* species which do not possess well defined sheath around its cells. With temperature there is no dispute about the internal environment experienced by the organism, since this is an environmental parameter which cannot be regulated, and therefore, a means of adaptation is to posses cellular components which are stable to temperature. A large number of enzymes from thermophiles have been found to be very thermostable (Zeikus *et al.*, 1991), and interestingly, these enzymes also have unusual stability to organic solvents, detergents, proteolytic agents, and pH extremes (Fontana, 1990). Thus thermophilic blue green algae are qualitatively different from the mesophiles. Certain organisms of the latter group possess the ability to adapt to high temperature and desiccation for brief periods only by developing special adaptive morphological features which very in response to the environmental conditions. To the contrary for the thermophiles, hot springs are their natural environments with relative constancy in their physico-chemical characteristics where the organisms have evolved to meet the environmental challenges of high temperatures, thus do not develop any such adaptive morphological features in their cells to survive at high temperature.

□□□

7

Summary

Survey of certain industrial waste water receiving areas (Co-operative Sugar factory and distillery, Aska; Orissa Sponge-iron limited, Palasponga; Fertilizer factory, Talcher) of Orissa State, India for the occurrence of various blue green algae forms has been presented. All these industries discharge large quantities of untreated or partially treated liquid wastes containing objectionable amounts of various organic wastes and inorganic salts to the surrounding area including the rice fields. A variety of blue green algal species, predominantly ensheathed forms, belonging to the genera *Chroococcus*, *Gloeothece*, *Gloeocapsa*, *Calothrix* and organisms of LPP group (*Lyngbya -Plectonema - Phormidium*) occur in the waste water receiving areas through out the year. These organisms also retained their sheath in the laboratory cultures.

Calothrix marchica Lemm. var *intermedia* Rao has been found to grow abundantly in soil receiving the effluent of the distillery. The effluent at low concentrations (1-10%, v/v) and at neutral pH increased the growth of the organism. The organism grew mixotrophically, photoheterotrophically as well as chemoheterotrophically in the medium supplemented with sucrose and lower concentration (10%, v/v) of neutralized distillery effluent; the latter containing 0.4 to 0.6% reducing sugars. The exogenous carbon sources were also utilized in certain specific biosynthetic process, for example in the synthesis of sheath layer around the trichome of the alga. Formation of sheath was not observed in the medium with mannose, pentoses and organic acids which did not support the heterotrophic growth.

Survival of two heterocystous blue green algae, *Calothrix castelli*, isolated from habitats polluted by the effluent of a sponge-iron factory and *Anabaena torulosa* from unpolluted fields, on various concentrations of the waste water of the industry was studied. *C. castelli*, which has a well defined sheath tolerated higher concentrations of the effluent in comparison to *A. torulosa*. Loss of sheath coupled with hair formation was observed in iron- deficient medium. Iron-rich culture favoured the sheath development. Further studies showed that *Gloeothece* PCC 6909 and *Calothrix marchica* which possess thick sheath around the cells/trichome tolerated higher concentrations of the waste water from the distilley, sponge-iron factory and fertilizer factory than the thin sheathed *Westiellopsis prolifica* from rice fields and *Anacystis nidulans* CCAP 1405/1 in laboratory culture. The extracelluar sheath isolated from *Gloeothece* PCC 6909 showed that it was composed of predominantly carbohydrate and protein. Quantitative differences in the carbohydrate and protein content of the sheath was observed between the organism grown with or without 10% (v/v) of the neutralized effluent of the distillery, sponge-iron and fertilizer factory.

The diluted effluent of the distillery (1-10%, v/v) supported seedling growth of the rice, *Oryza sativa* MTU-17. When 10% (v/v) neutralized distillery effluent and the blue green alga *Calothrix marchica* Lemm. var. *intermedia* Rao were supplemented in different combinations, they could significantly increase the vegetative growth and yield of the paddy. Outdoor cultivation of *C. marchica* singly or in mixed-culture with other nitrogen fixing blue green algae, *Anabaena* sp. and *Scytonema schmidlei* in presence of 10% (v/v) neutralized distillery effluent produced higher amount of algal biomass within a short time compared to the conventional autotrophic mode of biofertilizer inoculum production.

In addition to the studies on blue green algae in habitats polluted by industrial effluents, their occurrence, growth and nitrogen fixation in pesticide burdened rice field soils is also presented. The effect of four commonly used pesticides, Furadan (carbofuran, 3%), Sevin (carbaryl, 50% DB), Rogor (dimethoate, 30% EC) and Endotaf (endosulfan, 35% EC) on the algae of rice field soils was examined. The organochlorine, Endotaf was found to be more toxic than the organophosphate, Rogor and carbamate pesticides, Sevin and Furadan, in that sequence. The cyanophyceae members were more resistant than the chlorophycean and bacillariophycean members. *Lyngbya major, Gloeocapsa atrata, Calothrix marchica* and *Scytonema pascheri,* all of which have well-defined sheath occurred in the soil in the presence of 1000, 500 and 250 ppm of the carbamate, organophosphate and organochlorine pesticides respectively, where other blue green algal forms could not survive.

The effect of all these four pesticides on the survival, growth, photosynthetic oxygen evolution and nitrogen fixation of *Westiellopsis Prolifica* and *Calothrix marchica* in laboratory culture has been given. A progressive decrease in the growth, nitrogen fixation and rate of photosynthetic oxygen evolution of both the blue green algae with increasing concentrations of the pesticides in the medium was observed, however, *C. marchica* was always more tolerant than *W. prolifica* to increasing pesticide doses. The EC 50 values of the pesticides on growth of both the organisms showed that *C. marchica* possessing a distinct sheath around its trichome tolerated comparatively higher doses of the agrochemicals than the thin sheathed *W. prolifica*. These results of laboratory experiments correlated well with the findings on the reaction of blue green algae of rice field soils to pesticide application that the species occurring in soil possessing a sheath layer tolerated higher concentrations of the pesticides than the sheath less forms.

Epilithic blue green algae occurring on the exposed rock surfaces of temples and monuments from various regions of India and Nepal under temperature and desiccation stresses is also described. In all the 32 different temples surveyed, mostly blue green algal species occurred as blackish-brown crust/tuft on the temple walls even during summer months under extreme climatic conditions. Seven species of *Tolypothrix*, one species each of *Phormidium*, *Lyngbya*, *Gloeocapsopsis* and *Plectonema* appeared soon after wetting of the crusts/tufts collected from different locations. Thirty three species/strains of blue green algae belonging to *Gloeothece*, *Myxosarcina*, *Chroococcidiopsis*, *Plectonema*, *Nostoc*, *Calothrix*, *Chlorogloeopsis*, *Hapalosiphon* and *Fischerella* appeared in the enriched culture of the crusts/tufts alongwith the dominant species. The blue green algal forms which form predominant epilithic flora on the temples grew slowly, showed higher respiratory oxygen uptake rate than their corrresponding rate of photosynthesis and possess a well defined sheath around their cells/trichome. Most of these species occurred on many temples of different regions irrespective of the variation in their geographical location. All the 44 epilithic blue green algal isolates have been maintained in culture collection of Utkal University.

Of these several blue green algae, a single *Tolypothrix* species, *T. byssoidea* occurs predominantly on the rock surface of Sun temple, Konark, Jagannath temple, Puri and several temples of Bhubaneswar, in the east coast of Orissa state, India. The organism persists even during mid summer months forming blackish-brown crust when the temperature on the rock surfaces goes beyond 60° C coupled with high light intensity and extreme dryness. Treatment of the crusts with 2,3,5-triphenyl-tetrazolium chloride (TTC) showed that reduced activity was maintained in the algal cells even under extreme arid conditions. Photosynthesis and respiration started within 10 min of wetting and reached maximum activity in about 24 hours suggesting that the internal structure of the organism remained without damage even after exposure to strong solar insolation.

Absorption spectra of methanolic extracts of *Tolypothrix byssoidea* isolated from the rock surfaces of the temples and grown under fluorescent light showed prominent absorption at 260, 330 and 384 nm in the UV range of the spectrum. The organism was able to survive and retain the absorption peaks at 260 and 384 nm even after 24 h of UV-C irradiation. In contrast, a *Tolypothrix* sp. isolated from rice fields could not survived 30 min of similar UV irradiation and did not show absorption at 384 nm. The isolated sheath fraction of *T. byssoidea* absorbed strongly in the near- UV-blue region of the spectrum. The pigment absorbing at 384 nm reflects an adaptive strategy of terrestrial blue green algae to cope with damaging short-wavelength solar radiation.

The algal crust occurring on the exposed rock surface of the temples over-produce specific proteins when exposed to UV radiation, synthesized two HMWs proteins and counteracted the UV damage. Upon exposure to heat, few polypeptides were repressed showing their adverse effect on the survival of the organism in the crust which was previously subjected to higher temperature in desiccated state for longer duration. However, in the crust collected from natural habitat, several LMWs proteins (10.5, 13, 25 kDa), Wsp 39 kDa, additional proteins under 42 kDa group (43, 49 kDa), a chaperonin (58 kDa) and stress-90 group (84 kDa) were prominent which possibly was the reason for the adaptation of the organism there-in to the extreme temperature and desiccation. Unlike the modification pattern of protein synthesis in the blue green alga inhabiting the crust, the organism grown in culture could tolerate heat by synthesizing two HMWs, though three proteins under LMWs group were repressed. Several proteins of diverse groups were also repressed in the organism upon exposure to UV-C, however, was counteracted by induction of three new polypeptides under chaperonin and HMWs group and over producing one 41 kDa protein.

The blue green alga *Aulosira* sp. occurred as brownish wolly crust on the rice field soils after harvest of kharif crop and remained in the fallow rice fields through out the summer season exposed to intense solar radiation also possessed a thick, coloured sheath layer around its trichome. Unlike the alga in the epilithic crusts, the organism in the curst on the soils absorbed water rapidly and became saturated within 10 min of wetting. Appearance of metabolic activities and reappearance of the filamentous structure of the organisms inhabiting the crust on the soil was faster than the species in the epilithic crusts on the temple walls showing that more severe the drought condition a blue green alga was subjected to, longer time period it required for resumption of its structure and metabolic functions. Further, the *Aulosira* sp. synthesized scytonemin only when desiccated and simultaneously exposed to bright sunlight which was lost within 48 h of wetting, and was never formed in culture under fluorescent light. This shows that extracelluar scytonemin is formed only in certain blue green algal species possessing a sheath around the cell wall and occurring in habitats exposed to strong insolation coupled with desiccation.

Ecophysiology of blue green algae inhabiting at high temperature in the thermal springs of Orissa state has been presented. The temperature of the thermal springs: Taptapani in Ganjam district and Autri in Puri district, remained almost constant, 44° C and 55° C respectively. The water of both the springs were alkaline with a pH between 9.2 to 9.6. No carbon dioxide and only traces of nitrate, nitrite and phosphate was detected. A number of blue green algae and diatoms, and few zooplankton were recorded in these hot water springs. The quantity of plankton was maximal in the main tank of Taptapani and only in the overflows of Autri. None of the organisms except the blue green algae *Mastigocladus laminosus* and *Oscillateria teribriformis*, and the diatoms, *Navicula* sp. and *Cyclotella* sp. were found in the main tank of Autri at 55° C.

Maximum growth and branching development of *Mastigocladus laminosus* was observed at 45° C in the laboratory culture. Growth of the organism was also higher at pH 8.5 though pH 9.5 of the medium was found to be most effective for maximum frequency of the branching in the thallus. Photosynthetic oxygen evolution in the cells of the organism was also active at high temperatures and the optimum temperature agreed with that of its growth. The growth and photosynthetic activities were low at sub-optimal temperatures and irreversibly inactivated at temperature above 60° C. The blue green algae flourishing in these thermal springs at higher temperature, however, did not possess a sheath layer around their cell/trichome. This shows that thermophilic blue green algae are qualitatively different from the mesophiles, the former groups being evolved in their specialized natural habitat of great antiquity meeting the challenges of high temperature.

Literature Cited

Adhikary, S.P. 1983. Growth measurements by monitoring light scattering of a filamentous blue green alga which does not give uniform and stable suspension in culture vessels. Z. *Allg. Mikrobiol.* **23**: 475-483.

Adhikary, S.P. 1987. Growth response of *Calothrix marchica* Lemm. var. *intermedia* Rao to exogenous organic substrates and distillery effluent in light and dark. *J. Basic Microbiol.* **27**: 475-481.

Adhikary, S.P. 1996. Ecology of freshwater and terrestrial cyanobacteria. *J. Sci. Ind. Research*, **55**: 753-762.

Adhikary, S.P. 1998. Polysaccharides from mucilaginous envelope layers of cyanobacteria and their ecological significance. *J. Sci. Ind. Research*, **57**: 454-466.

Adhikary, S.P. and Pattnaik, H. 1979. Growth response of *Westiellopsis prolifica* Janet to organic substrates in light and dark. *Hydrobiol.* **67**: 241-247.

Adhikary, S.P. and Sahu, J. 1988. Production of sheath of *Calothrix marchica* Lemm. in photo-and chemoheterotrophic culture. *Curr. Sci.* **57**: 91-93.

Adhikary, S.P. and Sahu, J. 2000. Studies on the establishment and Nitrogenase activity of inoculated cyanobacteria in the field and their effect on the yield of rice. *Oryza*, **37**: 39-43.

Adhikary, S.P. and Satapathy, D.P. 1996. *Tolypothrix byssoidea* (Cyanophyceae/ cyanobacteria) from temple rock surfaces of coastal Orissa, India. *Nova Hedwigia*, **62**: 419-427.

Adhikary, S.P.; Dash, P. and Pattnaik, H. 1994. Effect of the carbamate insecticide sevin on *Anabaena* sp. and *Westiellopsis prolifica*. *Acta Microbiol. Hungaric.* **31**: 335-338.

Adhikary, S.P., Weckesser, J., Jürgens, U.J., Golecki, J.R. and Borowiak, D. 1986. Isolation and chemical characterization of the sheath from the cyomobacterium *chroococcus minutus* SAG B.41.79. *J. Gen. Microbiol.*, **132**: 2595-2599.

Agarwal, M. and Kumar, H.D. 1978. Physico-chemical and phycological assessment of two mercury polluted environments. *I.J. Env. Health*, **20**: 141-155.

Ahmad, M.H. and Venkataraman, G.S. 1973. Tolerance of *Aulosira fertilissima* to pesticides. *Curr. Sci.* **42**: 108-114.

Ahluwalia, A.S. and Arora, A. 1986. Effect of an industrial effluent on the growth of blue-green and green algae. In: *Green Vegetation Research*, Eds. Singh, N; Dukstra, J and Venkataraman, L.V., Soc. Green Veg. Res., Varanasi, pp. 19-31.

Ahluwalia, A.S.; Kaur, M. and Dua, S. 1989. Physico-chemical characteristics and effect of some industrial effluents on the growth of a green alga *Scenedesmus* sp. *Ind. J. Env. Health*, **31**: 112-114.

Albertano, P. 1993. Epilithic algal communities in hypogean environments. *Giorn Bot. Italy*, **127**: 385-395.

Albertano, P. and Grilli Caiola, M. 1989. A hypogean algal association. *Braun Blanquetia*, **3**: 287-292.

Albertano, P.; Kovacik, L. and Grilli Caiola, M. 1994. Preliminary investigation on epilithic cyanophytes from a Roman Necropolis. *Arch. Hydrobiol. Algological Studs.***75**: 71-74.

Allen, M. B. 1953. The thermophilic aerobic spore forming bacteria. *Bacteriol. Rev.* **17**: 125-173.

Allen, M.B. and Arnon, D.I. 1995. Studies on nitrogen fixing blue green algae. I. Growth and nitrogen fixation by *Anabaena cylindrica* Lemm. *Pl. physiol.* **30**: 366-372.

Amemiya, Y. and Nakayama, O. 1984. The chemical composition and metal absorption capacity of the sheath material isolated from *Microcystis*, cyanobacteria. *Jpn. J. Limnol.* **45**: 187-193.

Anagnostidis, K. 1977. Die cyanophyceen als Begleit organismen der Salz-, Süss- und Thermalwasser-Sulphureten Griechenlands. *Schweiz. Z. Hydrol.* **39**: 130-133.

Anagnostidis, K.; Economou-Amilli, A. and Roussomoustakaki, M. 1983. Epilithic and chasmolithic microflora (Cyanophyta/Bacillariophyta) from marbles of the parthenon (Acropolis-Athens, Greece). *Nova Hedwigia*, **38**: 227-277.

Anand, N. 1989. *A handbook of Blue-green Algae*. Bisen Singh Mahendra Pal Singh, Dehradun, India, p. 79.

Anand, N. and Veerappan, B. 1980. Effect of pesticides and fungicides on blue green algae. *Phykos*, **19**: 210-212.

Andrews, J.N. 1991. Radioactivity and dissolved gases in the thermal waters of Bath. In: *The Hot springs of Bath*, Ed. Kellaway, G.A., Bath city council, U.K., pp. 157-170.

A.P.H.A. (American Public Health Association), American water works Association and water pollution control Federation, 1965, 1971, 1989. *Standard methods for examination of water and waste water*, APHA, New York, USA.

Apte, S.K.; Fernandes, T.A.; Iyer, V. and Alahari, A. 1997. Molecular basis of tolerance to salinity and drought stresses in photosynthetic nitrogen-fixing cyanobacteria. In: *Plant Molecular Biology and Biotechnology*. Eds. Tewari, K.K. and Singhal, G.S., Narosa Publishing House, New Delhi, pp. 259-268.

Ashkey, J.; Rusforth, S.R. and Johanesen, F.R. 1985. Soil algae of cryptogamic crusts from the Uintah Basin, Utah, U.S.A. *Great Basin Nat.* **45**: 432-442.

Bell, R.A.; Athey, P.V. and Sommerfeld, M.R., 1986. Cryptoendolithic algal communities of the Colorado plateau. *J. Phycol.* **22**: 429-435.

Belnap, J. and Gardner, J.S. 1993. Soil microstructure in soils of Colorado plateau: the role of cyanobacterium *Microcoleus vaginatus*. *Great Basin Naturalist*, **53**: 40-47.

Bewley, J.D. 1979. Physiological aspects of desiccation tolerance. *Ann. Rev. Plant Physiol.* **30**: 195-238.

Bhagwat, A.A and Apte, S.K. 1989. Comparative analysis of proteins induced by heat shock, salinity, and osmotic stress in the nitrogen-fixing cyanobacterium *Anabaena* sp. strain L-31. *J. Bacteriol.* **171**: 5187-5189.

Bharadwaja, Y. 1934. The taxonomy of *Scytonema* and *Tolypothrix*. *Rev. Algol.* 7: 149-178.

Bishop, C.T.; Adams, G.A. and Hughs, E.O. 1954. A polysaccharide from the blue green alga *Anabaena cylindrica*. *Can. J. Microbiol.* **11**: 877-885.

Blondin, P.A.; Kirby, R.J. and Barnum, S.R. 1993. The heat shock response and acquired thermotolerance in three strains of cyanobacteria. *Curr. Microbiol.* **26**: 79-84.

Böhm, G.A.; Pfleiderer, N.; Böger, P. and Scherer, S. 1995. Structure of a novel oligosaccharide-mycosporine amino acid UV A/B absorbing pigment from the terrestrial cyanobacterium *Nostoc commune. J. Biol. Chem.* **270**: 8536-8539.

Borbely, G., Suranyi, G., Korcz, A. and Palifi, Z. 1985. The effect of heat shock on protein synthesis in the cyanobacterium *Synechococcus* sp. strain PCC 6301. *J. Bacteriol.* **161**: 1125-1130.

Bosch, T.C.G.; Krylow, S.M.; Bode, H.R. and Steele, R.E. 1988. Thermotolerance and synthesis of heat shock proteins: these responses are present in *Hydra attenuaia* but absent in *Hydra oligactis. Proc. Natl. Acad. Sci. USA*, **85**: 7927-7931. Brock, T.D. 1967a. Microorganisms adopted to high temperature. *Nature*, Lond. **214**:882-885.

Brock, T.D. 1967b. Life at high temperatures. *Science*, N.Y. **158**: 1012-1019.Brock, T.D. 1973. Lower pH limit for the existance of blue green algae: evolution and ecological implications. *Science*, **179**: 480-483.

Brock, T.D. 1975. Effect of water potential on a *Microcoleus* (Cyanophyceae) from a desert crust. *J. Phycol.* **11**: 316-320.

Brock, T.D. 1978. *Thermophilic microorganisms and life at high temperatures.* Springer, New York, p. 633.

Brock, T.D. and Brock, M.L. 1966. Temperature optima for algal development in Yellowstone and Iceland hot springs. *Nature*, **29**: 733-734.

Brock, T.D. and Brock, M.L. 1967. The hot springs of Furnas valley, Azores. *Int. Rev. Ges. Hydrobiol.* **52**: 545-558.

Brock, T.D., Brock, M.L., Bott, T.L. and Edwards, M.R. 1971. Microbiol life at 90° C: the sulfur bacteria of Boulder spring. *J. Bact.* **107**: 303-314.

Brown, T.E. and Richardson, F.L. 1968. The effect of growth environment on the physiology of algae: light intensity. *J. Phycol.* **4**: 38-54.

Buckley, C.E. and Hughton, J.A., 1976. A study of the effects of near UV-radiation on the pigmentation of the blue green alga *Gloeocapsa alpicola. Arch. Microbiol.* **107**: 93-97.

Büdel, B. and Wessels, D.C.J. 1991. Rock inhabiting blue-green algae/cyanobacteria from hot arid regions. *Arch. Hydrobiol, Algological Studs.* **64**: 385-398.

Cameron, R.E. 1960. Communities of soil algae occurring in the Sonaran desert in Arizona. *J. Arizona Acad. Sci.* **3**: 85-88.

Campbell, S.E. 1979. Soil stabilization by a prokaryotic desert crust : implications for precambrian land biota. *Origin of life*, **9**: 335-348.

Castenholz, R.W. 1969. Thermophilic blue-green algae and the thermal environment. *Bacteriol. Rev.* **33**: 476-504.

Castenholz, R.W. 1970. Laboratory culture of thermophilic cyanophytes. *Schweiz. Z. Hydrol.* **32**: 538-551.

Castenholz, R.W. 1973. Ecology of blue-green algae in hot springs. In: *The Biology of Blue-Green Algae*. Eds. Carr, N.G and Whitton, B.A., Blackwell, Oxford, pp. 379-414.

Castenholz, R.W. 1978. The biogeography of hot spring algae through enrichment cultures. *Mitt. Int. Verein. Limnol.* **21**: 296-315.

Castenholz, R.W. and Wickstrom, C.E. 1975. Thermal streams. In: *River Eology*. Ed. Whitton, B.A., Blackwell, Oxford, pp. 264-284.

Chada, A., Pandey, D.C. and Tiwari, G.L. 1980. Investigations on the algae inhabiting the walls of buildings. General survey. In: *Proc. Natl. Workshop Algal Systems*. Eds. Seshadri, C.V., Thomas, S. and Jeejibai, N, Indian Institute of Technology, Madras, pp. 238-240.

Chitnis, P.R. and Nelson, N. 1991. Molecular cloning of the genes encoding two chaperone proteins of the cyanobacterium *Synechocystis* PCC 6803. *J. Biol. Chem.* **266**: 58-65.

Collyer, D.M. and Fogg, G.E. 1955. Studies on fat accumulation by algae. *J. Expt. Bot.* **17**: 256-275.

Copeland, B.K. 1967. Biological and physiological basis of indicator organisms. In: *Pollution and marine ecology*. Eds. Olson, T.A. and Burgess, F.J., Inter Science Pubishers, Willey, New York, pp. 285-288.

Costerton, J.W.; Cheng, K.J.; Geesey, G.G.; Ladd, T.I.; Nickel, J.C.; Das-Gupta, M. and Marrie, T.J. 1987. Bacterial biofilm in nature and disease. *Ann. Rev. Microbiol.* **41**: 435-464.

Cox, E.R. and Hightower, J. 1972. Some corticolus algae of Mcmmin country, Tennessee, USA. *J. Phycol.* **8**: 203-205.

Craig, E.A. 1985. The heat shock response. *CRC Crit. Rev. Biochem.* **18**: 239-280.

Craig, E.A., Ingolia, T.D. and Manseau, L.J. 1983. Expression of *Drosophila* heat shock cognate genes during heat-shock and development. *Dev. Biol.* **99**: 418-422.

Crowe, J.H., Crowe, L.M., Carpenter, J.F. and Wistrom, C.A. 1987. Stabilization of dry phospholipid bilayers and proteins by sugars. *Biochem. J.* **242**: 1-10.

Danin, A. 1983. Weathering of limestone in Jerusalem by cyanobacteria. *Z. Geomorphol.* **27**: 413-421.

Danin, A., Bar-or, Y, Dor, I. and Ysraeli, T. 1989. The role of cyanobacteria in stabilization of sand dunes in southern Israel. *Ecol. Mediterranean*. **15**: 55-64.

Danin, A. and Garty, J. 1983. Distribution of cyanobacteria and lichens on hillside of the Negev highlands and their impact on biogenic weathering. *Z. Geomorph*. **27**:423-444.

Das, A.K. and Mishra, P.K. 1996. Changes in pigment and protein content of *Westiellopsis prolifica*, a blue green alga grown in paper mill waste water. *Micorbios*, **85**: 257-266.

Das, B. and Singh, P.K. 1977. Effect of weedicide 2,4. dichloro-phenoxy acetic acid on growth and nitrogen fixation of blue green algae *Anabaenopsis raciborskii*. *Arch. Environ. Contam. Toxicol*. **5**: 437-445.

Das, M.K. and Adhikary, S.P. 1996. Toxicity of three pesticides to several rice field cyanobacteria. *Trop. Agric. (Trinidad)*. **73**: 156-158, 1966.

Das, S.C.; Mandal, B. and Mandal, L.N. 1991. Effect of growth and subsequent decomposition of blue-green algae on the transformation of iron and manganese in submerged soils. *Plant and soil*, **138**: 75-84.

Dailva, E.J.; Henriksson, L.E. and Henriksson, E. 1975. Effect of pesticides on blue-green algae and nitrogen fixation. *Arch. Environ. Contam. Toxicol*. **3**: 193-204.

Davies, B.H. 1976. Carotenoids. In: *Chemistry and Biochemistry of Plant Pigments*. Academic Press, London, pp. 149-155.

De, P.K. 1939. The role of blue green algae in nitrogen fixation in rice fields. *Proc. Roy. Soc. Lond*. **127**: 121-134.

Desikachary, T.V. 1959. *Cyanophta*, ICAR monograph on Algae, ICAR, New Delhi, p.686.

De Winder, B.; Matthijis, H.C.P. and Mur, L.R. 1989a. The role of water retaining substrate on the photosynthetic response of three drought tolerant prototrophic microorganisms isolated from a terrestrial habitat. *Arch. Microbiol*. **152**: 458-462.

De Winder, B.D.; Pluis, J.; Reus, L.D. and Mur, L.R. 1989b. Characterisation of a cyanobacterial algal dune crust in the coastal dunes of the Netherlands. In: *Microbial mats : Physiological ecology of benthic microbial communities*. Eds. Choen, Y. and Rosenberg, E., American Society of Microbiol, Washington, pp. 77-83.

Dikshit, G. and Tiwari, G.L. 1992. Effect of pesticides on nitrogen fixing cyanobacteria. In: *Cyanobacterial Nitrogen Fixation*. Ed. Kaushik, B.D., Ass. Pub. Comp. Co., New Delhi, pp. 495-500.

Donker, V.A. and Häder, D.P. 1995. Protective strategies of several cyanobacteria against solar radiation. *J. Plant Physiol.* **145**: 750-755.

Donna R.H., Hladum, S.L., Scherer, S. and Potts, M. 1994. Water stress proteins of *Nostoc commune* (cyanobacteria) are secreted with UV-A/B pigments and associate with 1,4- B-D-xylanxylano hydrolase activity. *J. Biol. Chem.* **269**: 7726-7734.

Drouet, F. 1938. Myxophyceae of the Yale North Indian Expedition. Collected by G.E. Hutinson. *Trans Amm. Microbiol.* Soc. **57**: 127-131.

Drews, G. and Weckesser, J. 1982. Function, structure and composition of cell walls and external layers. In. *The Biology of Cyanobacteria.* Eds. Carr. N.G. and Whitton, B.A., Blackwell, oxford, pp. 333-357.

Dulieu, D.; Gaston, A. and Darley, J. 1977. La dégradation des Pâturages de La region N' Djamena en relation evec la presence de cyanophycees psamnophiles. *Rev. Elev. Med. Vet Pays Trop.* **30**: 181-190.

Dunn, J.H. and Wolk, C.P. 1970. Composition of cellular envelopes of *Anabaena cylindrica. J. Bacteriol.* **103**: 153-158.

Dupuy, P.; Trotet, G. and Grossini, F. 1976. Protection des monuments contre les cyoncphycees en milieu abrite et humide. In: *Science, Technology and European Cultural Heritage.* Proc. Sympos. Bologna, Italy. Eds. Bear, N.S., Sabbioni, C and Sors, A.I., Butterworth- Heinemann, Oxford, pp. 205-221.

Edmunds, W.M. and Miles, D.L. 1991. Geochemistry of Bath thermal waters. In: *The thermal springs of Bath.* Ed. Kellaway, G.A., Bath city council, UK, pp. 143-156.

Elchhron, G.L. 1975. Active sites of biological macromolecules and their interaction with heavy metals. In: *Ecological Toxicology Research.* Ed. McIntyre, A.D. and Mills, C.F., Plenum Press, New York.

Fitzerald, 1971. Bioassay analysis of nutrient availability. Report: *Water, air and waste chemistry.* American chemical society, Los Angles, pp. 1-18.

Fernandes, T.A.; Iyer, V. and Apte, S.K. 1993. Differential responses of nitrogen-fixing cyanobacteria to salinity and osmatic stresses. *Appl. Environ. Microbiol.* **59**: 899-904.

Fitz, S.; Fitz-Ulrich, E.; Frenzel, G.; Krüger, R. and Kühn, H. 1987. *Die Einwirkumg von luft-verunreinigurgan auf ausgewählte Kunstwerke Mittelalterlicher Glasmalerei,* E. Schmidt, Berlin.

Fogg, G.E. 1949. Growth and heterocyst production in *Anabaena cylindrica* Lemm. II. In relation to carbon and nitrogen metabolism. *Ann. Bot.* **13**: 241-259.

Fogg, G.E. 1971. Recycling through algae. *Proc. Roy. Soc. Land. B.* **179**: 201-207.

Fogg, G.E.; Stewart, W.D.P.; Fay, P. and Walsby, A.E. 1973. *The Blue Green Algae,* Academic Press, New York and London, p. 459.

Fontana, A. 1990. How nature engineers protein thermostability. In: *Life under Extreme Conditions.* Ed. Diprisco, G., Springer Verlag, Heidelberg, Germany, pp. 89-113.

Forster, S.M. and Nicholson, T.H. 1981. Microbial aggregation in maritime dune succession. *Soil. Biol. Biochem.* **13**: 205-208.

Friedmann, E.I. 1962. The ecology of the atmophytic nitrate-alga *Chrococcidiopsis kashaii* Friedmann. *Arch. Microbiol.* **42**: 42-45.

Friedmann, E.I. 1971. Light and scanning electron microscopy of the endolithic desert algal habitat. *Phycologia,* **10**: 411-428.

Friedmann, E.I. 1972. Ecology of lithophytic algal habitats in middle Eastern and North American deserts. In: *Ecophysiological foundation of ecosystems productivity in avid zone.* Ed: Rodin, L.E., Nanka, USSR Acad Sci. Leningrad, pp. 182-185.

Friedmann, E.I. 1977. Microorganisms in Antarctic desert rocks from dry valleys and Dufek Massif Antarctic *J.U.S.,* **12**: 26-30.

Friedmann, E.I. 1982. Endolithic microorganisms in the Antarctic cold desert. *Science,* **215**: 1045-1053.

Friedmann, E.I., Lipkin, Y. and Ocampo-Paus, R. 1967. Deseart algae of the Negev (Israel). *Phycologia,* **6**: 185-200.

Friedmann, E.I. and Ocampo-Friedmann, R. 1976. Endolithic blue-green algae in the dry valleys : primary producers in the Antarctic desert ecosystem. *Science,* **193**: 1247-1249.

Friedmann, E.I. and Ocampo-Friedmann, R. 1984. Endolithic microorganisms in extreme dry environments : analysis of a lithobiontic microbial habitat. In: *Current Prospectives in Microbial Ecology.* Eds. Klug, M.J. and Reddy, C.A., American Soc. Microbiol., Washington, pp. 177-185.

Fritsch, F.E. 1922. The moisture relations of terrestrial algae. I. Some general observations and experiments. *Ann. Bot.* **36**: 1-20.

Fritsch, F.E. and Haines, F.M. 1923. The moisture relations of terrestrial algae. II. The changes during exposure to drought and treatment with hypertonic solutions. *Ann. Bot.* **37**: 683-728.

Fritz-Sheridan, R.P. 1988. Physiological ecology of nitrogen fixing blue green algae crusts in the upper sub-alpine life zone. *J. Phycol.* **24**: 302-309.

Fujita, Y., Ohki, K. and Murakami, A. 1987. Chromatic regulation of photosystem composition in the cyanobacterial photosynthetic system : Kinetic relationship between change of photosystem composition and cell proliferation. *Plant Cell Physiol.* **28**: 227-234.

Galambos, J.T. 1967. The reaction of carbazole with carbohydrates. *Anal. biochem.* **19**: 119-132.

Gangawane, L.V. 1979. Tolerance of Thimet by nitrogen-fixing blue-green algae. *Pesticides*, **13**: 33-34.

Gangawane, L.V. and Kulkarni, L. 1979. Tolerance of certain fungicides by nitrogen fixing blue-green algae and their side effect on rice cultivers. *Pesticides*, **13**: 37-39.

Gangawane, L.V. and Saler, R.S. 1979. Tolerance of certain fungicides by nitrogen fixing blue green algae. *Curr. Sci.* **48**: 306-308.

Gantt, E. 1994. Supramolocular membranae organization. In. *Molecular Biology of Cyanobacteria.* Ed. Bryant, D.A., Kluwer, Dordrecht, pp. 119-138.

Garcia-Pichel, F. and Castenholz, R.W. 1991. Characterization and biological implications of scytonemin, a cyanobacterial sheath pigment. *J. Phycol.* **27**: 395-409.

Garcia-Pichel, F. and Castenholz, R.W. 1993. Occurrence of UV-absorbing, mycosporine-like compounds among cyanobacterial isolates and an estimate of their screening capacity. *Appl. Environ. Microbiol.* **59**: 163-169.

Garcia-Pichel, F.; Sherry, N.D. and Castenholz, R.W. 1992. Evidence for a UV sunscreen role of the extracelluar pigment scytonemin in the terrestrial cyanobacterium *Chlorogloeopsis* sp. *Photochem. Photobiol.* **56**: 17-23.

Garcia-Pichel, F.; Wingard, C.E. and Castenholz, R.W. 1993. Evidence regarding UV sunscreen role of mycosporine-like compound in the cyanobacterium *Gloeocapsa* sp. *Applied. Environ. Microbiol.* **59**: 170-176.

Geitler, L. 1932. Cyanophyceae. In: *Kryptogamen flora von Deutschland, Österreich and der Schweiz.* Ed. Rabenhorst, L., Akadem. Verlagsges Leipzig, p. 1196.

Gething, M.J. and Sambrock, J. 1992. Protein folding in the cell. *Nature,* **355**: 33-34.

Giacobini, C.; Andreoli, C.; Casadoro, G.; Fumanti, B.; Lanzara, P. and Rascio, N. 1979. Una caratteristica alterazione delle murature edegli intonaci. In. *Proc. Int. Cong. Deterioramentoe conservazione della pietra,* Attidel. Ed. Baden, B., Venezia, Padova, pp. 289-299.

Gloaguen, V.; Morvan, H. and Hoffmann, L. 1995. Released and capsular polysaccharides of Oscillatoriaceae (cyanophyceae, cyanobacteria). *Algol. Studs*. **78**: 53-69.

Goedheer, J.C. and Kleienchammans, J.W. 1997. Growth and chemical composition of blue green alga *Anacystis nidulans* cultured at high light intensities. *Acta. Bot. Neerl*. **26**: 273-284.

Golecki, J.R. 1977. Studies on ultrastrcture and composition of the cell wall of the cyanobacterium *Anacystis nidulans. Arch. Microbiol*. **114**: 35-41.

Gonzalves, E. 1947. The algae flora of the hot springs of Vjreswari near Bombay. *J. Univ. Bombay*. **16**: 22-27.

Gour, J.P. and Kumar, H.D. 1986. Effects of oil refinery effluents on *Sclenastrum capricornutum* Printz. *Int. Revue ges. Hydrobiol*. **71**: 271-281.

Graebner, P. 1910. Plantzen leben auf die Dümen. In: *Dünenbuch*. Eds. Sloger, F.; Graebner, P.; Thienemann, J.; Speiser, P. and Schulze, F.W.O. Enke verlag, stuttgart, pp. 183-296.

Goyal, S.K. 1993. Algal biofertilizer for vital soil and free nitrogen. Proc. *Ind. Natl. Sci. Acad*. **B 59**: 295-302.

Herbert, D.; Phipps, P.J. and Strange, R.E, 1971. Chemical analysis of microbial cells. In: *Methods in Microbiology*. vol. 5B. Eds. Norris, J.R. and Ribbons, D.W., Academic Press, New York and London, pp. 210-344.

Hewitt, E.J. 1963. Mineral nutrition in plants in culture media. In: *Plant physiology*. Ed. Steward, F.C., Academic Press, New York, pp. 97-123.

Hoffmann, L. 1989. Algae of terrestrial habitats. *Bot. Rev*. **55**: 77-105.Hoffmann, L. and Demoulin, V. 1985. Morphological variability of some species of Scytonemataceae (cyanophyceae) under different culture conditions. *Bull. Soc. Roy. Bot. Belg*. **118**: 189-197.

Hough, L., Jones, J.K.N. and Wadman, W.H. 1952. An investigation of the polysaccharide component of certain freshwater algae. *J. Chem. Soc*. 3393-3399.

Hustedt, F. 1930. Bacillariophycae, Die Kieselagen. In: *Rabenhorsts Kryptogamen Flora* 7. Leipzig, Germany.

Jaag, O. 1995. Untersuchungen uber die vegetation und Biologie der nackten Gestins in den Alpen, im Jura und im Schweizerischen Mittleland. Beitr. Kryptogamenfl. Schweiz. **9**: 1-560.

Jana, B.B. 1978. The plankton ecology of some thermal springs in West Bengal, India. *Hydrobiol*. **61**: 135-143.

Jha, M. and Kumar, H.D. 1986. Cyanobacterial flora of two hotsprings of Rajgir, Bihar. In: *Green vegetation and leaf protein Research*. Eds. Singh, N.; Dijkstra, J. and Venkataraman, L.V., BHU, Varanasi, pp. 33-40.

Johansen, J.R. 1993. Cryptogamic crusts of semiarid and arid lands of North America. *J. Phycol.* **29**: 140-147.

Johansen, J.R.; Ashley, J. and Rayburn, W.R. 1993. The effects of range fire on soil algae crusts in semiarid shrub-steppe of the lower Columbia basin and their subsequent recovery. *Great Basin Nat.* **53**: 64-69.

Jørgensen, B.B. and Nelson, D.C. 1988. Bacterial zonation, photosynthesis and spectral light distribution in hot spring microbial mats of Iceland. *Mirob. Ecol.* **16**: 133-147.

Kannayan, S. 1978. Effect of insecticides on the growth of blue green algae. In: *Pesticide Residues in the Environment in India*. Eds. Edwards, C.A.; Veeresh, G.K. and Krueger, H.R. FAO/UNDP, Bangalore, pp. 492-495.

Kar, G.K.; Mishra, P.C.; Dash, M.C. and Das, R.C. 1987. Pollution studies in river Ib. *Ind. J. Environ Health*, **29**: 322-329.

Kar, S. and Singh, P.K. 1977. Effect of pH, light intensity and population on the toxicity of the pesticide Carbofuran to the blue green alga *Nostoc muscorum*. *Microbios.* **21**: 177-184.

Kar, S. and Singh, P.K. 1978. Toxicity of carbofuran to the blue green alga *Nostoc muscorum*. *Bull. Env. Contam. Toxicol.* **22**: 707-714.

Kar, S. and Singh, P.K. 1979a. Detoxification of pesticides Carbofuran and Hexachlorocyclohexane to the blue green alga *Nostoc muscorum*. *Microbios Lett.* **10**: 111-114.

Kar, S. and Singh, P.K. 1979b. Effect of nutrients on the toxicity of pesticides Carbofuran and Hexachloro-cyclohexare to blue green alga *Nostoc muscorum*. *Z. Allg. Microbiol.* **19**: 467-472.

Karsten, U. and Garcia-Pichel, F. 1996. Carotenocids and mycosporine amino acid like compounds in members of the genus *Microcoleus* (cyanobacteria): a chemosystematic study. *System. Appl. Microbiol.* **19**: 285-294.

Kashyap, A.K. and Pandey, K.D. 1982. Inhibitory effects of rice field herbicide Mechete on *Anabaena doliolum* Bhardwaja and protection by nitrogen sources. *Z. Pflan. Physiol.* **107**: 339-345.

Kaur, M., Dua, S. and Ahluwalia, A.S. 1993. Effect of effluent of paper factories on the growth of *Scenedesmus*. *Phykos*, **32**: 57-63.

Kaushik, B.D. and Venkataraman, G.S. 1983. Response of cyanobacterial nitrogen fixation to insecticides. *Curr. Sci.* **52**: 321-323.

Khalil, K., Chaporkar, C.B. and Gangawane, L.V. 1980. Tolerance of blue-green algae to herbicides. In. *Proc. Workshop Algal Systems.* Ind. Soc. Biotech. IIT New Delhi, pp. 36-39.

Kliner, E.F. and Harper, K.T., 1977. Soil properties in relation to cryptogamic ground cover in canyonlands National Park. *J. Range. Manage.* **30**: 202-205.

Kirtikar, K.R. 1886. A new species of algae *Conferva thermalis* Birdwoodii. *J. Bom. Nat. Hist. Soc.* **1**: 135-138.

Kolte, S.O. and Goyal, S.K. 1992. On the effect of herbicides on growth and nitrogen fixation by cyanobacteria. *Acta Bot. Indica,* **20**: 225-229.

Komárek, J. 1993. Validation of the genera *Gloeocapsopsis* and *Asterocapsa* (cyanoprocaryota) with regard to species from Japan, Mexico and Himalayas. *Bull. Natl. Sci. Mus. Tokyo,* Ser B. **19**: 19-37.

Kratz, W.A. and Myers, J. 1955. Nutrition and growth of several blue green algae. *Ann. J. Bot.* **42**: 282-287.

Kumar, A.; Tyagi, M.B.; Srinivas, G.; Singh, N.; Kumar, H.D.; Sinha, R.P. and Häder, D.P. 1996. UV-B shielding role of $FeCl_3$ and certain cyanobacterial pigments. *Photochem. Photobiol.* **64**: 321-325.

Kumar, H.D. and Sharma, V. 1974. Experimental studies on the uptake of mineral nutrients by algae isolated from polluted habitats. *Biochem. Physiol. Pflanzen.* **166**: 221-232.

Krumbein, W.E. 1972. Rôle des microorganisms dan la genése, la diagenése et la dégradation des roches en place. *Rev. Ecol. Biol. Sol.* **9**: 283-319.

Krumbein, W.E. 1988. Biotransformations in monuments-a sociobiological study. *Durable Build. Mat.* **5**: 359-382.

Krumbein, W.E. and Lange, C. 1978. Deccay of plaster, paintings and wall materials of the interior of buildings via microbiol activity. In: *Environmental Biogeochemisty and Geomicrobiology.* Ed. Krumbein, W.E., Ann Arbor, pp. 687-697.

Landry, J.; Bernier, D.; Chretien, P.; Nicole, L.M.; Tanguay, R.M. and Marceau, N. 1982. Synthesis and degradation of heat shock proteins during development and deccay of thermotolerance. *Cancer Res.* **42**: 2457-2463.

Lange, G.J.; Kidron, B.; Büdel, A.; Meyer, E.; Kilian, E. and Abliovich, A. 1992. Taxonomic composition and photosynthetic characteristics of the biological soil crusts covering sand dunes in the Western Negev desert. *Funct. Ecol.* **6**: 519-527.

Laszlo, A. and Li, G.C. 1985. Heat resistant variants of chinese hamster fibroblasts altered in expression of heat shock protein. *Proc. Natl. Acad. Sci.* USA, **82**: 8029-8033.

Levi, Y.; Berner, T. and Cohen, Y. 1981. CO_2 exchange and growth rate of the loess soil crust algae in the Neger desert of Israel. In: *Developments in avid zone ecology and environmental quality.* Ed. Shuval, H., Balaban L.C.S., Philadelphia P.A., pp. 43-48.

Levine, E. and Thiel, T. 1987. UV-inducible DNA repair in the cyanobacterium *Anabaena* spp. *J. Bacteriol.* **169**: 3988-3993.

Lindquist, S. 1986. The heat shock response. *Ann. Rev. Biochem.* **55**: 1151-1191.

Lindquist, S. and Craig, E.A. 1988. The heat shock proteins. *Ann. Rev. Genet.* **22**: 631-677.

Ljungdahl, L.G. and Sherod, D. 1976. Proteins from thermophilic microorganisms. In: *Extreme Environments.* Ed. Heinrich, M.R., Academic press, New York, pp. 147-187.

Lowe, S.E.; Jain, M.K. and Zeikus, J.G. 1993. Biology, ecology and biotechnological applications of anaerobic bacteria adopted to environmental stresses in temperature, pH, salinity or substrates. *Microbial Rev.* **57**: 451-509.

Lowry, O.H.; Roberts, N.R.; Leiner, K.Y.; Wu, M.L. and Farr, M.L. 1954. The quantitative histochemistry of brain. I. chemical methods. *J. Biol. Chem.* **207**: 1-17.

Mackinney, G. 1941. Absorption of light by chlorophyll solutions. *J. Biol. Chem.* 140:315-322.Manoharan, C. and Subramanian, G. 1993. Feasibility studies on using cyanobacteria in Ossein effluent treatment. Ind. *J. Environ. Health*, **35**: 88-96.

Marathe, K.V. and Chaudhuri, P.R. 1975. An example of algae as pioneer in the lithosphere and their role in rock corrosion. *J. Ecol.* **63**: 65-70.

Margulis, B.A.; Antroprova, O.Y. and Kharazova, A.D. 1989. 70 kDa heat shock proteins from mollusc and human cells have common structural and functional domins. *Comp. Biochem. Physiol.* **94**: 621-642.

Marimoto, R.I.; Tissieres, A. and Georgopoulos, C. 1990. *Stress proteins in biology and medicine.* Cold Spring Harbour press, New York.Mc Cann, A.E. and Cullimore, D.R. 1979. Influence of pesticides on the soil algal flora. *Residue Rev.* **72**: 1-31.

Mc Donald, G.C.; Spear, R.D.; Lavin, P.J. and Clesari, N.L. 1970. Kinetics of algal growth in Austere media. In: *Properties and Products of Algae.* Ed. Jajic, J.E., Plenum press, New York and London, pp. 97-105.

Mehta, U.B. and Vaidya, B.S. 1978. Cellular and extracellular polysaccharides of the blue green alga *Nostoc*. *J. Exp. Bot.* **29**: 1423-1430.

Megharaj, M.; Venkateswarlu, K. and Rao, A.S. 1986. Effect of Monocrotophos and Quinalphous soil on algae. *Environ. Poll.* **40**: 121-126.

Megharaj, M., Venkateswarlu, K. and Rao, A.S. 1987. Influence of Cypermethrin and Fenvalorate on a green alga and three cyanobacteria isolated from soil. *Ecotox. Environ. Saf.* **14**: 142-146.

Megharaj, M.; Venkateswarlu, K. and Rao, A.S. 1988. Tolerance of algal population in rice soils to Carbofuran application. *Curr. Sci.* **57**: 100-102.

Metting, B. 1988. Microalgae in agriculture. In: *Microalgal Biotechnology*. Eds. Borowitzka, M.A. and Borowizka, L.J., Cambridge Univ. press, Cambridge, pp. 288-304.

Mittler, R. and Tel-or, E. 1991. Oxidative stress response in the unicellular cyanobacterium *Syrechococcus* PCC 7942. *Free Rad. Res. Comm.* **12**: 845-850.

Mishra, A.K.; Pandey, A.B. and Kumar, H.D. 1989. Effects of three pesticides on MSX-induced ammonia photoproduction by the cyanobacteria *Nostoc linckia*. *Ecotox. Env. Saf.* **18**: 145-148.

Moore, B.G. and Tischer, R.G. 1965. Biosynthesis of extracellular polysaccharides by the blue green alga *Anabaena flos-aquae*. *Can. J. Microbiol.* **11**: 877-885.

Nagpal, V. and Goyal, S.K. 1992. Growth responses of cyanobacteria to herbicides. *Acta. Bot. Ind.* **20**: 173-176.

Nayak, H.; Sahu, J.K. and Adhikary, S.P. 1996. Blue green algae of rice fields of Orissa. II. Growth and nitrogen fixing potential. *Phykos*, **35**: 111-118.

Oldham, T. and Oldham, R.D. 1882. The thermal springs of India. *Mem. Geol. Survey India*, **19**: 99-161.

Ortega Calvo, J.J.; Hernandez-Marine, M. and Saiz-Jimenz, C. 1993. Isolation and characterization of epilithic chlorophytes and cyanobacteria from two Spanish Cathedrals (Salamanca and Toledo). *Nova Hedwigia*, **57**: 239-253.

Ortega Calvo, J.J.; Arino, X; Stal, L.J; and Saiz-Jimenez, C. 1994. Cyanobacterial sulfate accumulation from black crust of a historic building. *Geomicrobiol. J.* **12**: 15-22.

Oswald, W.J. and Golueke, C.G. 1971. Harvesting and processing of waste grown micro algae. In: *Algae, Man and Environment*. Ed. Jackson, D.F., Syracure Univ. press, New York.

Pabbi, S. and Vaishya, A.K. 1992. Effect of insecticides on cyanobacterial growth and nitrogen fixation. In: *Cyanobacterial Nitrogen Fixation*. Ed. Kaushik, B.D., Ass. Publ. Co., New Delhi, pp. 489-493.

Padhy, R.N. 1985. Cyanobacteria and pesticides. *Residue Rev.* **95**: 1-44.

Painter, T.J. 1983. Algal polysaccharides. In: *The Polysaccharides*. vol.2. Ed. Aspinall, G.O., Academic press, New York.

Palmer, C.M. 1969. A composite rating of algal tolerating organic pollution. *J. Phycol.* **5**: 78-82.

Palmer, C.M. 1980. *Algae and water pollution*. Castle House publications Ltd., England, pp. 31-78.

Pandey, K.D. and Kashap, A.K 1986. Differential sensitivity of three cyanocobacteria to the rice field herbicide Mechate. *J. Basic Microbiol.* **26**: 421-428.

Parker, D.L.; Schram, B.R.; Plude, J.L. and Moore, R.E. 1996. Effect of metal cations on the viscosity of a pectin like capsular polysaccharide from the cyanobacterium *Microcystis flos-aquae* C3-40. *Appl. Environ. Microbiol.* **62**: 1208-1213.

Patridge, S.M. 1949. Paper chromatography of sugars *Nature* (Lond), **164**: 443.

Pattnaik, H. 1964. *Studies on nitrogen fixation by Westiellopsis prolifica Janet*. Ph.D. thesis. University of London, U.K.

Peary, J. and Castenholz, R.W. 1964. Temperature strains of a thermophilic blue green alga. *Nature (Lond.),* **202**: 720-721.

Peat, A.; Powell, N. and Potts, M. 1988. Ultrastructural analysis of the rehydration of desiccated *Nostoc commune* HUN (Cyanobacteria) with particular reference to the immunolabelling of Nif H. *Protoplasma*, **146**: 72-80.

Pentecost, A. 1985. Investigation of variation in heterocyst numbers, sheath development and false branching in natural populations of Scytonemataceae (Cyanobacteria). *Arch. Hydrobiol.* **102**: 343-353.

Pietriini, A.M.; Ricci, S.; Bartolini, M. and Giulani, M.R. 1985. A reddish color alteration caused by algae on stone works. In: *Proc. Int. Cong. Deterioration and conservation of stone*. Ed. Felix, G., Romandes press, Lausanne, pp. 653-663.

Pipe, A.E. 1992. Pesticide effects on soil algae and cyanobacteria. *Review Environ. Contam. Toxicol.* **127** 95-169.

Plesset, J.; Palm, C. and McLaughlin, C.S. 1982. Induction of shock proteins and thermotolerace by ethanol in *Saccharomyces cerevisiae. Biochem. Biophys. Res. Comm.* **108**: 1340-1345.

Plinius, S. 1944. *Natural History*, Harvard Univ. Press, Cambridge.

Plude, J.L.; Parker, D.L.; Schommer, O.J.; Timmerman, R.J.; Hagstrom, S.A.; Joers, J.M. and Hnasko, R. 1991. Chemical characterization of polysaccharide from the slime layer of the cyanobacterium *Microcystis flos-aquae* C3-40. *Appl. Env. Microbiol.* **57**: 1969-1700.

Potts, M. and Friedmann, E.I. 1981. Effects of water stress on cryptoendolithic cyanobacteria from hot desert rocks. *Arch. Microbiol.* **130**: 257-271.

Potts, M.; Ocampo-Friedmann, R.; Bowman, M.A. and Tözun, B. 1983. *Chrococcus* S 24 and *chroococcus* N 41 (cyanobacteria): morphological, biochemical and genetic characterization and effects of water stress on ultrastructure. *Arch. Microbiol.* **135**: 81-90.

Prasad, B.N. and Srivastava, P.N. (1965). Thermal algae from Himalayan hot springs. *Proc. Natl. Acad. Sci.* India. **31 B**: 45-53.

Prescott, G.W. 1951. *Algae of western Greatlake area*. Cran Book, London, U.K.

Pritzer, M.; Weckesser, J. and Jürgens, U.J. 1989. Sheath and outermembrane components from the cyanobacterium *Fischerella* sp. PCC. 7414. *Arch. Microbiol.* **153**: 7-11.

Proteau, P.J.; Gerwick, W.H.; Garcia-Pichel, F. and Castenholz, R.W. 1993. The structures of scytonemin, a ultraviolet sunscreen pigment from the sheaths of cyanobacteria. *Experientia,* **49**: 825-829.

Rana, B.C.; Gopal, T. and Kumar, H.D. 1971. Studies on biological effects of industrial wastes on the growth of algae. *Environ. Health,* **13**: 138-143.

Rai, L.C. and Kumar, H.D. 1976a. Algal growth as a means of evaluation of nutrient status of the effluent of a fertilizer factory near Shahupuri, Varanasi. *Tropical Ecol.* **17**: 50-56.

Rai, L.C. and Kumar, H.D. 1976b. Nutrient uptake by *Chlorella vulgaris* and *Anacystis nidulans* isolated from the effluent of a fertilizer factory. *Ind. J. Ecol.* **3**: 63-69.

Rath, B. and Adhikary, S.P. 1984. Relative tolerance of several nitrogen-fixing cyanobacteria to commercial grade furadan (Carbofuran, 3% G). *Ind. J. Expt. Biol.* **32**: 213-215.

Reddy, T.R.K.; Rao, j.C.S. and Radhakrishnan, T.M. 1983. Toxicity of oil refinery effluent on physiological responses of algae. *Phykos.* **22**: 86-93.

Reynolds, E.S. 1963. The use of lead citrate at high pH as an electron opaque stain in electron microscopy. *J. Cell. Biol.* **17**: 208-212.

Rippka, R. 1972. Photoheterotrophy and chemoheterotrophy among unicellular blue-green algae. *Arch. Microbiol.* **87**: 93-98.

Rippka, R.; Deruelles, J.; Waterbury, J.B., Herdman, M.A. and Stanier, R.Y. 1979. Generic assignments, strain histories and properties of pure cultures of cyanobacteria. *J. Gen. Microbiol.* **111**: 1-61.

Rodgers, G.A.; Bergman, D.; Henriksson, E. and Urdis, M. 1979. Utilization of blue green algae as biofertilizer. *Plant Soil,* **52**: 99-107.

Rodriguez-Lopez, M., 1965. Utilization of sugars by *Chlorella* under various conditions. *J. Gen. Microbiol.* **43**: 139-143.

Roger, P.A. 1988. Reconsidering the utilization of blue green algae in wetland rice cultivation. In: *Biological Nitrogen Fixation Associated with Rice production.* Eds. Dutta. S.K. and Sloger, C., Oxford and IBH publ. Co., New Delhi. pp. 119-137.

Roger, P.A. and Kulasooriya, S.A. 1980. *Blue green algae and Rice,* International Rice Research Institute, Los Banos, Philippines.

Roger, P.A. and Reynand, P.A. 1982. Free living blue green algae in tropical soils. In: *Microbiology of Tropical Soils and Plant Productivity.* Eds. Dommergues, Y.R. and Diem, H.G., Martinus Nijhoffl, Dr. W. Junk, The Hague, The Netherlands, pp. 147-168.

Roger, P.A.; Santiago-Ardales, S.; Reddy, P.M. and Watanabe, I. 1987. The abundance of heterocystous blue green algae in rice soils and inocula used for application in rice fields. *Biol. Fertil. Soils,* **5**: 98-105.

Rosen, H. 1957. A modified ninhydrin colorimetric analysis for amino acids. *Arch. Biochem. Biophys.* **67**: 10-15.

Round, F.E. 1973. *The biology of the algae.* 2nd Ed., Arnold press, London, p. 247.

Roy, A.; Tripathy, P. and Adhikary, S.P. 1997. Epilithic blue-green algae/ cyanobacteria from temples of India and Nepal. Presence of UV sunscreen pigments. *Algol. Studs.* **86**: 147-161, 1997.

Roychoudhury, P. and Kaushik, B.D. 1986. Response of cyanobacterial growth and nitrogen fixation to herbicides. *Phykos,* **25**: 36-43.

Saha, S.K. and Dutta Munshi, J. 1983. Algal flora of Sita Kund and Bhimbandh thermal springs of Bihar. *Biol. Bull. Ind.* **5**: 257-261.

Saha, S.K.; Dutta Munshi, J.; Dutta Munshi J.S. and Sarkar, H.L. 1978. Limnobiotic survey of thermal springs of Bhimbandh. *Geobios,* **5**: 205-207.

Sangar, V.K. and Dugan, P.R. 1972. Polysaccharide produced by *Anacystis nidulans* : its ecological implications. *Appl. Microbiol.* **14**: 732-734.

Scherer, S. and Potts, M. 1989. Novel water stress protein from a desiccation tolerant cyanobacterium. *J. Biol. Chem.* **264**: 12546-12553.

Scherer, S., Chen, T.N. and Böger, P. 1986. Recovery of adenine nucleotide pools in terrestrial blue green algae after prolonged drought periods. *Oecologia,* **68**: 585-588.

Scherer, S.; Ernst, A.; Chen, T.W. and Böger, P. 1984. Rewetting of drought resistant blue-green algae: time course of water uptake and reappearance of respiration, photosynthesis and nitrogen fixation. *Oecologia,* **62**: 418-423.

Scherer, S.; Chen, T.W. and Böger, P. 1988. A new UV A/B protecting pigment in the terrestrial cyanobacterium *Nostoc commune. Plant Physiol.* **88**: 1055-1057.

Schlesinger, M.J. 1986. Heat shock proteins: the search for functions. *J. Cell. Biol.* **103**: 321-327.

Schlesinger, M.J. and Hershako, A. 1988. *The Ubiquitin system.* Cold Spring Harbour press, Cold Spring Harbor, New York.

Schopf, J.W. 1968. Microflora of the Bitter springs formation, late Precambrian, central Australia. *J. Palentol.* **42**: 651-688.

Schopf, J.W. 1975. Precambrian paleobiology : problems and prospectives. *Ann. Rev. Earth Plamet. Sci.* **3**: 213-249.

Shephard, K.L. 1987. Evaporation of water from the mucilage of a gelatinous algal community. *Br. Phycol. J.* **22**: 181-185.

Shubert, L.E. and Starks, T.L. 1978. Algal succession on orphaned coal mined soils. In: Ecology of Coal Resource Development. Ed. Wali, M.K., Pergamon press, New York, pp. 661-669.

Shields, L.M. and Drouet, F. 1962. Distribution of terrestrial algae within the Nevada test site. *Amer J. Bot.* **49**: 547-554.

Singh, A.K. and Gour, J.P. 1988. Effect of Assam crude oil on photosynthesis and associated electron transport system in *Anabaena doliolum. Bull. Env. Contam. Toxicol.* **41**: 776-780.

Singh, A.K. and Kumar, H.D. 1991. Inhibitory effect of petroleum oil on photosynthetic electron transport system in the cyanobacterium *Anabaena doliolum. Bull. Env. Contam. Toxicol.* **47**: 890-895.

Singh, A.L., Singh, P.K. and Singh, P.L. 1988. Effects of different herbicides on the *Azolla* and blue green algae biofertiization of rice. *J. Agric. Sci. Camb.* **111**: 451-458.

Singh, M. 1993. Studies on weathering of Kailashnath temple, Kancheepuram. *Curr. Sci.* **64**: 559-565.

Singh, H.N. and Vaishyampayan, A. 1978. Biological effect of rice field herbicide Mechate on various strains of the nitrogen fixing blue green alga *Nostoc muscorum*. *Env. Expt. Bot.* **18**: 87-94.

Singh, L.J. and Tiwari, D.N. 1988. Some important parameters in the evaluation of herbicide toxicity in diazotrophic cyanobacteria. *J. Appl. Bacteriol.* **64**: 365-376.

Singh, P.K. 1973. Effect of pesticides on blue-green algae. *Arch. Microbiol.* **89**: 317-320.

Singh, P.K. 1985. Nitrogen fixation by blue green algae in paddy fields. In: *Rice Research in India*. Eds. Jaiswal, P.L.; Wadhwani, A.M.; Singh, R.; Chhabra, N.N. and Chhabra, N., ICAR, New Delhi, pp. 344-362.

Singh, R.N. 1961. *Role of blue green algae in nitrogen economy of Indian agriculture*. ICAR, New Delhi, p. 175.

Sinha, R.P.; Lebert, M.; Kumar, A.; Kumar, H.D. and Häder, P. 1995. Spectroscopic and biochemical analysis of UV effects on phycobiliproteins of *Anabaena* sp. and *Nostoc carnium*. *Bot. Acta.* **108**: 87-97.

Singleton, R. Jr. and Amelunxen, R.E. 1973. Proteins from thermophilic microorganisms. *Bacteriol. Rev.* **37**: 320-342.

Sinha, B.D. and Chaubey, L.M. 1986. Cyanobacterial flora of Bhimbandh hot spring. In: *Green vegetation and leaf protein research*. Eds. Singh, N.; Dijkstra, J. and Venkataraman, L.V., BHV, Varanasi, pp. 1-9.

Skarpe, C. and Henriksson, E. 1987. Nitrogen fixation by cyanobacterial crusts and by associative symbiotic bacteria in Western Kalahari, Botswana. *Arid. Soil Res.* Rehabil. **1**: 55-59.

Smith, A.J. 1982. Modes of cyanobacterial carbon metabolism. In. *Biology of Cyanobacteria*. Eds. Carr, N.G. and Whitton, B.A., Blackwell, Oxford, pp. 47-85.

Smith, G.M. 1950. *The freshwater algae of United States*. 2nd Ed., Mc Graw Hill, New York.

Soliman, A.I.; Dessouki, S.A.S.; Mansour, F.A. and Hussain, M.H. 1993. Effect of Batachlor on growth, pigmentation and nitrogen fixation. *Nostoc khilmani* and *Anabaena oscillarioides*. *Phykos*, **32**: 85-93.

Somasekhar, R.K. and Ramaswamy, S.N. 1983. Algal indicators of paper mill waste water. *Phykos*, **22**: 161-166.

Souza, K.S.; Kostiw, L.L. and Tyson, B.J. 1974. Alternations in normal fattyacid composition in a temperature sensitive-mutant of a thermophilic *Bacillus*. *Arch. Microbiol.* **97**: 89-102.

Stanier, R.Y. 1973. Autotrophy and heterotrophy in unicellular blue green algae. In: *The Biology of Cyanobacteria*. Eds. Carr, N.G. and Whitton, B.A., Black well, oxford, pp. 501-508.

Stanier, R.Y. and Cohen-Bazire, G. 1977. Phototrophic procaryotes, the cyanobacteria. *Ann. Rev. Microbiol.* **31**: 225-274.

Starks, T.L.; Shubert, L.E. and Trainor, F.R. 1981. Ecology of soil algae. *Phycologia*, **20**: 65-80.

Starmarch, K. 1966. *Cyanophyta* (Sinice-Glacofity). In: Flora Slodkowodna Polski, Polska Akad, Warszawa. 2: 1-807.

Stewart, W.D.P. 1974. *Algal Physiology and Biochemistry*, Blackwell, Oxford.

Stewart, W.D.P., 1980. Some aspects of structure and function in nitrogen fixing cyanobacteria. *Ann. Rev. Microbiol.* **34**: 497-536.

Stokes, J.L. 1940. The influence of environmental factors upon the development of algae and other microorganisms in soil. *Soil Science*, **49**: 171-184.

Stratton, G.W. 1987. The effects of pesticides and heavy metals towards phototrophic microorganisms. *Rev. Environ. Toxicol.* **3**: 71-147.

Stockner, J.G. 1967. Observations of thermophilic algal communities in Mount Rainer and Yellowstone National Parks. *Limno lcoeanogr.* **12**: 13-17.

Subramanian, G.; Krishnamurthi, M.S. and Lakshmiprabha, S. 1987. Influence of some insecticides on photosynthetic oxygen evolution in the cyanobacterium *Anabaena*. *Curr. Sci.* **56**: 549-550.

Subramanian, G.; Sekhar, S. and Sampoornam, S. 1994. Biodegradation and utilization of organophosphorus pesticides by cyanobacteria. *Int. Biodet. Biodeg.* **33**: 129-143.

Swaminathan, M.S. 1984. Rice. *Scientific American*, **250**: 62-71.

Tandon, R.S.; Lal, R. and Rao, V.V.S.N. 1988. Interaction of Endosulfan and Malathion with blue-green algae *Anabaena* sp. and *Aulosira fertilissima*. *Env. Poll.* **52**: 1-9.

Tarar, J.L. and Shawale, T.H. 1984. Studies on the effects of some fungicides on soil algae of paddy fields. *Phykos*, **23**: 191-201.

Tease, B.E. and Walker, R.W. 1987. Comparative composition of the sheath of the cyanobacterium *Gloeothece* ATCC 27152 culture with or without combined nitrogen. *J. Gen. Microbiol.* **133**: 3331-3339.

Thomas, J. and David, K.A.V. 1971. Studies on the physiology of heterocyst production in the nitrogen fixing blue green alga *Anabaena* sp. L-31 in continuous culture. *J. Gen. Microbiol.* **66**: 127-131.

198   *Blue Green Algae : Survival Strategies in Diverse Environments*

Thomas, J. and Gonzalves, E.A. 1965. Thermal algae from Western India. *Hydrobiol.* **26**: 21-65.

Thomas, P.S. and Shanmugasundaram, S. 1986. Interaction of agrochemicals with the cyanobacterium *Anacystis nidulans* and patterns of growth. *Microbios Lett.* **33**: 115-120.

Tomaselli, L. and Giovannetti, L. 1993. Survival of diazotrophic cyanobacteria in soil. *World J. Microbiol. Biotechnol.* **9**: 113-116.

Tomaselli, L.; Margheri, M.C. and Florenzano, G. 1979. Indagine sperimentale sul ruolodei cianobatterie delle microalghe nel deterioramento di monumenti ed affreschi. In: *Proc. Int. Cong. Deterioramento conservazionae della pietra. Attlidel.* Ed. Baden, B., Venezia, Padova, Italy, pp. 313-325.

Trainor, F. 1985. Survival of algae in a desiccated soil : a 25 year study. *Phycologia*, **24**: 79-82.

Triol, A.C.; Roger, P.A. and Watanabe, I. 1982. Fate of nitrogen from a blue-green alga in a flooded rice soil. *Soil. Sci. Plant. Nutr.* **28**: 559-569.

Tripathy, I.C.; Ayyappan, S.; Pandey, B.K. and Adhikary, S.P. 1990. Isolation and growth of local heterocystous cyanobacteria for using as biofertilizer. In: *Cyanobacterial Nitrogen Fixation.* Ed. Kaushik, B.D., Associated publ. company Ltd., New Delhi, pp. 275-278.

Tripathy, P.; Roy, A. and Adhikary, S.P. 1997. Survey of epilithic blue-green algae (cyanobacteria) from temples of India and Nepal. *Algol. Studs.* **87**: 43-57.

Tripathy,P., Roy,A., Anand,N. and Adhikary,S.P. 1999. Blue green algal flora on the rock surface of temples and monuments. Fedds Repertorium,110:132-144.

Tripathy, S.N. and Talpasayi, E.R.S. 1980. Sulphydryl and survival of subaerial blue green algae. *Curr. Sci.* **49**: 131-132.

Tyagi, R.; Kumar, H.D.; Vyas, D. and Kumar, A. 1991. Effects of ultraviolet-B radiation on growth, pigmentation, NaH ^{14}Co$_3$ uptake and nitrogen metabolism in *Nostoc muscorum*. In: *Global Climatic Changes on Photosynthesis and Plant Productivity.* Indo-US Workshop, New Delhi, Oxford and IBH, pp. 109-124.

Uhlinger, D.J. and White, D.C. 1983. Relationship between physiological status and formation of extracellular polysaccharide glycocalyx in *Pseudomonas atlantica*. *Appl. Environ. Microbiol.* **45**: 64-70.

Van den Ancker, J.A.M.; Jungerius, P.D. and Mur, L.R. 1985. The role of algae in the stabilization of coastal dune blow outs. *Earth surf. proc. Landf.* **10**: 189-192.

Van der Molen, J.M.; Garty, J.; Aardema, W. and Krumbein, W.E. 1980. Growth control of algae and cyanobacteria on historical monuments by mobile UV unit (MU VU). *Studs. Conservation*, **25**: 71-77.

Vasistha, P.C. 1968. Thermal cyanophyceae of India. *Phykos*, **7**: 198-241.

Venkataraman, G.S. 1969. *The cultivation of Algae.* ICAR manograph, New Delhi p. 319.

Venkataraman, G.S. 1972. *Algal Biofertilizer and rice cultivation.* Today and Tomorrows printers and publishers, New Delhi p.75.

Venkatarman, G.S. 1975. The role of blue-green algae in tropical rice cultivation. In: *Nitrogen Fixation by Free living Microorganisms.* Ed. Stewart, W.D.P., Cambridge Univ. press, pp. 207-218.

Venkataraman, G.S. 1978. *Algal biofertilizer for rice.* Informative mannual, Shiv Narain printers, New Delhi.

Venkataraman, G.S. 1981. *Blue-green algae for rice production-a mannual for its promotion.* FAO soil bulletin, Rome, Italy, p. 102.

Venkataraman, G.S. and Neelakanthan, S. 1967. Effect of cellular constituents of the nitrogen fixing blue green algae *Cylindrospermum muscicola* on the root growth of rice seedlings. *J. Gen. Appl. Microbiol.* **13**: 53-61.

Venkataraman, G.S. and Rajyalaxmi, B. 1971. Tolerance of blue green algae to pesticides. *Curr. Sci.* **40**: 143-144.

Venkataraman, G.S. and Rajyalaxami, B. 1972. Relative tolerance of nitrogen fixing blue green algae to pesticides. *Ind. J. Agric. Sci.* **42**: 119-121.

Venu, P.; Kumar, V.; Sardana, R.K. and Bhasin, M.K. 1984. Indicatory and functional role of phytoplankton in the effluents of Rangpo distilleries of Sikkim, Himalayas. *Phykos*, **23**: 38-44.

Vestal, J.R. 1985. Biomass of the Cryptoendolithic microbiota from the Antarctic desert. *Appl. Environ. Microbiol.* **54**: 957-959.

Wang, W.S. and Tischer, R.G. 1973. Study of the extracellular polysaccharides produced by a blue-green alga, *Anabaena flosaquae* A 37. *Arch. Microbiol.* **91**: 77-81.

Watanabe, A. 1962. Effect of nitrogen fixing blue-green alga *Tolypothrix tenuis* on the nitrogenous fertility of paddy soil and on the crop yield of rice plant. *J. Gen. Appl. Microbiol.* **8**: 85-91.

Watson, M.L. 1958. Staining of tissue sections for electron microscopy with heavy metals. *J. Biophys. Biochem. Cytol.* **4**: 475-478.

Webb, R.; Reddy, K.J. and Sherman, L.A. 1990. Regulation and sequence of the *Synechococcus* strain PCC 7942 groEL operon, encoding a cyanobacterial chaperonin. *J. Bacterol.* **172**: 5079-5088.

Weckesser, J.; Broll, C.; Adhikary, S.P. and Jürgens, U.J. 1987. 2-O-methyl-D-Xylose containing sheath in the cyanobacterium *Gloeothece* sp. PCC 6501. *Arch. Microbiol.* **147**: 300-303.

Weekesser, J.; Hoffmann, K.; Jürgens, U.J.; Whitton, B.A. and Raffelsberger, B. 1988. Isolation and chemical analysis of the sheaths of the filamentous cyanobacteria *Calothrix parietina* and *C. scopulorum*. *J. Gen. Microbiol.* **134**: 629-634.

Wee, Y.C. 1980. Ecology of subaerial algae. In: *Phycotalk*, vol.2. Ed. Kumar, H.D., Rastogi publ., Meerut, pp. 245-261.

Wessels, D.C.J. and Büdel, B. 1995. Epilithic and cryptoendolithic cyanobacteria of Clarens sandstone cliffs in the Golden-gate highland national park. South Africa. *Bot. Acta.* **108**: 169-276.

Whitton, B.A. 1987. Survival and dormancy of blue-green algae. In: *Survival and Dormancy of Microorganisms*. Ed. Henis, Y., Wiley, New York, pp. 109-167.

Whitton, B.A. 1992. Diversity, ecology and taxonomy of the cyanobacteria. In: *Photosynthetic Procaryotes*. Eds. mann, N.H. and Carr, N.G., Plenum press, New York, pp. 1-51.

Whitton, B.A.; Donaldson, A. and Potts, M. 1979. Nitrogen fixation by *Nostoc* colonies on terrestrial environments of Aldabra Atoll, Indian Ocean. *Phycologia*, **18**: 278-287.

Whitton, B.A.; Rother, J.A. and Paul, A.R. 1988. Ecology of deepwater rice fields of Bangladesh. 2. chemistry of sites at Manikganj and Sonargaon. *Hydrobiol.* **169**: 21-30.

Wilkens, J. Schruber, U. and Vidaver, W. 1978. Chlorophyll fluorescence induction: an indicator of photosynthetic activity in marine algae undergoing desiccation. *Can. J. Bot.* **56**: 2787-2794.

Wright, S.J.L. 1978. Interactions of pesticides with microalgae. In: *Pesticide Microbiology*. Eds. Hill, I.R. and Wright, S.J.L., Academic press, New York, pp. 535-601.

Wright, S.J.L.; Stainthorpe, A.F. and Downs, J.D. 1977. Interactions of the herbicide Propanil and metabolite 3, 4-dichloro aniline with blue-green algae. *Acta. Phytopathol. Acad. Sci. Hungaric.* **12**: 51-60.

Yamaoka, T.; Satoh, K. and Katoh, S. 1978. Photosynthetic activities of a thermophilic blue-green alga. *Plant Cell Physiol.* **19**: 943-954.

Zajic, J.E. and Chiu, Y.S. 1970. Heterotrophic culture of algae. In: *Properties and Products of Algae*. Ed. Zajic, J.E., Plenum press, New York, pp. 1-47.

Zeikus, J.G.; Lee, C.; Lee, Y.E. and Saha, B.C. 1991. Thermostable saccharides : new sources, uses and biodesigns. In: *Enzymes in Biomass Conversion*. Eds. Leathan, C.F. and Himmel, M.E., American Chemical Soc., Washington D.C., pp. 36-51.

❑❑❑